WILD AND FREE

Dear Dad,
   Hope you enjoy these Antipodean monsters
on your birthday!
   Lots of love,
      Mark

WILD

AND FREE

AUSTRALIA'S NATURAL WORLD

THROUGH THE LENS OF NICHOLAS BIRKS

Reed New Holland

First published in Australia in 1998 by
New Holland Publishers Pty Ltd
Sydney • Melbourne • London • Cape Town

167 Drummond St, Carlton, VIC, 3053, Australia

3/2 Aquatic Drive, Frenchs Forest, NSW, 2086, Australia

24 Nutford Place, London, W1H 6DQ, United Kingdom

80 McKenzie Street, Cape Town, 8001, South Africa

National Library of Australia
  Cataloguing in publication data

Birks, Nicholas.
   Australia's natural world wild and free: through the lens of Nicholas Birks
   ISBN 1 86436 355 X

   1. Zoology – Australia – Pictorial works. 2. Birds –

   Australia – Pictorial works. 1 Title

591.994

Publishing General Manager: Jane Hazell
Publisher: Clare Coney
Editor: Margaret Barca
Designer: Andrew Cunningham
Cover Designer: Julie Rovis
Printer: Imago Productions, Singapore

I think I inherited my fascination with nature from my father, Norman Birks. When I was quite small he would pick me up in the car from home, during his lunch hour, and whisk me to the Adelaide Zoo. There we would sit eating tomato sandwiches, and watch African weaver-birds build their grass nests, or some other fascinating creature. Half an hour later I would be back home and Dad would be at the office again. I remember him taking me to Humbug Scrub where we lay on the grass watching eagles soaring above for hours. My sister, Wendy, and I spent many weekends driving around the country with Dad talking to woodcutters and farmers, trying to find the elusive pygmy-possum. Little did I realize that one day they would live, naturally, in my own garden.

My first real pet was a homing pigeon and I can remember thinking the homing instinct was a fable when he was finally released and took off for the big city — never to be seen again.

My father died when I was ten and he was only forty-three. My little brothers were only babies, but Mum, Wendy and I missed him terribly.

As a child, walking across Adelaide's parks I learned about the little falcons that would follow and swoop on any small bird flushed by my

Kangaroo Island is a haven for wildlife and the people who live there really appreciate it. Thanks to Mike Mckelvie for sharing 'his' island with me. He was the inspiration for photographing the ospreys at Ballast Head. The staff at the National Parks and Wildlife Service on the island really care about the environment. Terry Dennis has, in his spare time, completed a definitive study on ospreys and other work on sea-eagles. With his help I managed to photograph the osprey at incredibly close range in the wild.

Ian Falkenberg is another national parks' man whose research on birds of prey is of benefit in planning the rehabilitation of injured birds. He has kindly shared his knowledge of the peregrine over the years.

I thank Peter Hornsby for sharing his extensive knowledge of rock wallabies. He changed the way I think about wildlife with his wise views based on psychology.

The help and hospitality of friends in the outback has been boundless. I would like to thank Nick and Sue Shearer from Bulloo Creek, whose tolerance and interest in eagles is refreshing. They share, with many other graziers like John and Judy Caskey at 'Larloona' and Pam and the late Ted

ACKNOWLEDGEMENTS

progress through the grass. I remember fielding at fine leg in a school cricket match, having, I thought, done my bit by opening the bowling. I was disturbed by shouts of, "Birks, get the bloody ball", as I tried to fish some new variety of trapdoor-spider out of its hole with a sour-sob stem.

There are so many people I would like to thank for their hospitality and help over the years. These include many former staff of the Adelaide Zoo, where I spent all my spare time while at boarding school. Old Bert, Max and Colin allowed me to join in with the rat hunts in the cages. The present staff have also been very helpful. My thanks to Melville for his help with the ghost bat in flight.

I spent weeks trying to photograph a platypus at Healesville and would like to thank Diane Log, Fisk and Barry for their help. Kevin Mason assisted with photographing the lyrebird. Also thanks to Andre Vanparidon and Paul Harrison for their help.

I have been inspired by my friends David Hollands, Lindsay Cupper and Penny Olsen. All have written books, which required tremendous effort and resourcefulness. Lindsay and David have been very generous, sharing their ideas, ethics and equipment.

Rieck of 'Merty Merty', a concern for their environment. Mutooroo and Mulyungarie, where I was once a jackaroo, has always been like a second home to me. Past and present managers like Laurie Manson, David Harris, John Manning and Richard Glouchester all share an interest in the land and its welfare. Thanks to my old neighbours at Willalooka: Tim and Janette Mort, Gary and Marny Fenton, and Tony and Pam Davis. Thank you for letting me climb the boundary fence in pursuit of bird and animal, for whom fences are just for sitting on.

I have always wanted to fly like an eagle and fairly recently this was made possible by Bob Mossel, adventurer and bush pilot.

The star of my life has always been my brilliant wife, Prudy. Her kind and positive approach to life has been a boost to all who know her and in particular our children. Without her hard work, as a physiotherapist and wise head we might have ended up with all our eggs in one basket. Alice, Danny and Benjamin are all unique and beautiful. They are now pursuing their own careers. We've had some great times together and I apologize for the odd occasion that I've chosen to sit in a hide rather than bowl a cricket ball at them. I also apologize for, at times, batting and bowling for too long.

This is a natural history book with a difference. There are no posed, static shots, but page after page of superb photographs showing birds, animals, reptiles and insects in action: an osprey gliding to its nest, kangaroos 'mid hop' and speeding bats frozen in flight in an instant of time.

I first heard of 'Nicky' Birks as an athlete in the late 1950s, although he competed some years after I retired. People spoke of this country lad at the National Championships, who with a minimum of prior competition took out the Australian javelin throwing title. He continued to do this regularly for several years, and won medals at the Commonwealth Games in Perth 1962 and Jamaica in 1966. In between time he won three Australian decathlon titles — a most exacting event which involves ten separate disciplines.

To have achieved this level of performance while training under primitive conditions and without adequate equipment gives us a clue to the discipline, focus and resourcefulness of the man, qualities which he has applied to his other great love, natural history.

Nicholas Birks has been a farmer for most of his life and this has enabled him to see nature in a broader way than many city-based naturalists do.

The disappearance of older trees, especially those with breeding hollows, is of great concern because they are the homes of many species of birds, bats and mammals. It is interesting to note thirty per cent of his own property near Keith, South Australia, was retained in native vegetation and fenced out from livestock. Although trees may live for some years if grazing is permitted, the fragile ecosystem of uncleared country is soon destroyed, the understorey dying first and eventually many of the dominant trees.

Yet Nicholas' story has an optimistic tone and he writes enthusiastically of the 'revolution' in chemical spraying which now takes note of timing and application rates to reduce the effect of the sprays on valuable parasites and predators. He sees the continued planting of trees, especially plant corridors, as a way of enhancing species diversity on farmland, and he notes how farm dams have contributed to the expansion of the population of kangaroos and some species of birds. He points out that birds such as sulphur-crested cockatoos, galahs and crested pigeons, are all commoner now than thirty years ago.

Many nature photographers favour telephoto lenses which enable them

FOREWORD BY JOHN LANDY

He regrets that city-based people have become disengaged from the natural world, and believes more needs to be done to keep city people in touch with nature, 'wild and free'.

The book is not solely about native wild life, as it includes photographs of alien and unwanted bush inhabitants, such as foxes, cats and rabbits. Nicholas does not shy away from the fact that some past farming practices have caused widespread environmental degradation. At the time these practices were implemented the consequences, which usually take many years to manifest themselves, were not understood. Soil erosion, excessive land clearing and salination are all discussed in a balanced way by a person who at one and the same time has been an active farmer intent on preserving his pastures and making a profit and yet a committed conservationist.

He points out that much of the past excessive land clearing was driven by a desire to reduce harbour for rabbits, which continue to cause immense environmental damage and economic loss to the Australian farm community. Hopefully the Calicivirus, which has spread to most Australian states, will greatly reduce their numbers.

to take photographs at a distance without disturbing the subject. Nicholas does not belong to this school, and believes on the contrary that wide-angle lenses with short focal lengths provide pictures that show the bird or the animal and its natural environment. Most photographers would agree that it is ideal if you can get very close to your subject and not in any way disturb it, using a wide-angle lens to simultaneously provide close-up details and a panoramic view of the surroundings. The fact is, however, that this is hugely difficult and is in my view the most significant skill that Nicholas Birks brings to his work. He has devised ways to sit for hours so close to the birds that they are unaware or unafraid of his presence. Also he has been willing to experiment with state-of-the-art technology, like infra-red photography, which has allowed him to take extraordinary photographs of nocturnal birds and bats in flight.

Above all, this book is a testimony to the incredible patience and tenacity of Nicholas Birks the photographer, to countless hours spent in bird hides and shelters, waiting to catch that single moment, in flight or on the run, which best defines the species. I believe that *Wild and Free* opens our eyes to Australian wildlife in a way that no other book has yet achieved.

A former Commonwealth Games representative, Nicholas Birks was a finalist in 1958, won a bronze medal in 1962 and a silver medal in 1966. He won 11 Australian championships in javelin and three in decathlon, and set five British Commonwealth javelin records and held the Australian javelin record for 18 years. Because he lived in the country on sheep and cattle stations during those competitive years, his training was often done in difficult circumstances. He used unusual methods to train for the event, without actually being able to throw a javelin very often. On the credit side, the training was rarely boring or repetitive. With only one or two competitions in some years, he competed in 17 Australian javelin titles and only missed a place in one of them, his first.

Nicholas and his wife Prudy owned a farm at Keith, in South Australia, for thirty years, where they brought up their three children: Alice, Daniel and Benjamin. They took up a block of land that was mostly scrub and carved out a sheep and cattle property of which they are very proud. Thirty per cent of the land was set aside as nature reserves and livestock excluded. During those years on the land one of his greatest loves was studying and photographing animal behaviour.

NICHOLAS BIRKS ARPS

Photography has always been a part of his life and wildlife has held a constant fascination. He used innovative methods to photograph wildlife, and eventually won the 1990 *Australian Magazine*/Nikon Wildlife Photographer of the Year Award.

He has provided photographs for books and magazine articles, as well as his own *Wildflight* range of postcards. He photographs a range of wildlife from spiders and insects to bats and birds in flight but his specialty has been birds of prey. The book *Australian Birds of Prey*, by Penny Olsen, for which he supplied two-thirds of the photographs, has recently won the Whitiley Prize for the best natural history book of the year.

At present, Nicholas is working on photographing Australian wildlife in action in outback landscapes, to expand his range of *Wildflight* cards.

Photography is a form of art. Composing a picture using light and colour takes some of the same skills an artist uses to compose work. It is not just a matter of recognising a potential photograph when you see it in nature. Often conceiving an image comes through imagination. Then it's a matter of setting out to achieve it. With wildlife photography, the ability to understand and anticipate the behaviour of the subject allows you to position the camera to take best advantage of the light before the action happens. Certainly there is craft involved with the use of equipment, as for an artist with a paintbrush, and a lot of practice is required for both. Many artists are great illustrators and capture every detail but, like much of John Gould's work, miss the expressions and light that make some photographs live.

A story is always in the back of my mind when taking photographs. Many subjects in nature create their own artistic forms as a frozen or moving image. The fantastic shapes formed by bats in flight are a delight. It's not always possible to anticipate how they will turn out on film, as the whole process takes place in total darkness and in the twinkling of an eye.

INTRODUCTION

The image of a bat on the wing scooping water, its reflection beneath, with droplets like diamonds spilling from its tongue, is tantalising. Another image of a bat carrying a beetle by its hind leg in through the shearing shed door was a revelation. These demonstrate the co-ordination and control a bat has while flying, as well as telling us something about their behaviour. Photographs showing the interactions within one species and between different species make a fascinating study.

You don't need to be a member of a conservation group to be aware of the problems facing many species. I have been a farmer for over thirty years and look back on my own mistakes with some sadness and regret. No matter where you look, from the deserts to the rainforest and wetlands, it's possible to see the effect of man. People make blunders that not only ruin things for many animals but can eventually ruin their own living as well. Sadly we don't consult our elders and if we do, we take only the advice we want to believe.

Feral animals and even native animals that are strangers to parts of Australia have been introduced over the years. Fish like golden perch have recently been released in watercourses for 'sport', but the fate of the many species of small native fish in those waters was not considered. It is sickening to see European carp in so many of our streams.

I have spent most of my life observing nature and learning the habits of mammals and birds, particularly birds of prey. I try to work away from nest sites as much as possible as some birds are very sensitive. I encourage them to come close enough for me to use a wide-angle lens, which allows the subject to be part of the environment. This can take months or even years to achieve. My aim is to photograph a complete natural history of all the birds I've studied. The period after fledging of the young, when they are learning to fend for themselves, is particularly fascinating.

One thing that stands out in most landscape photography is the lack of animal life. This is due to the small aperture and slow shutter speed used in taking that type of photograph. Even if a bird flew across the scene, it would not register on the film.

Stalking wildlife with a long telephoto lens is not my style. The resulting photographs contain birds and mammals with frightened expressions, looking back over their shoulders, and it is difficult to get the subject facing or moving in towards the camera.

Using a normal or wide-angle lens has an advantage over a telephoto one because it does not have to be moved to follow the progress of the subject. The slightest movement of a long lens will frighten most birds. The use of sound-deadening material to reduce the noise of the camera shutter is important at close range. I tend to concentrate on one subject in an area, which often leads to others in the vicinity. The hide is usually situated away from nests. Often the birds need no food, just a high perch from which to keep watch.

For many years I consistently fed a wild pair of peregrines and was able to observe much of their natural hunting technique from a tower fifteen metres high, placed in open country. In the autumn and early winter, after they had driven off their young from the previous season, the pair hunted together and would often sit on top of the hide, using it as an observation perch from which to hunt.

By climbing into the hide very early in the morning I beat their arrival, which was often before sunrise. The pair would sit only centimetres above my head preening and shuffling about. Sometimes I could actually feel their needle-sharp talons through the canvas on my skull. On winter Saturday mornings racing pigeons sometimes flew through on their way from the south-east to Adelaide or Murray Bridge. The peregrines had usually fed before the pigeons passed over, and ignored them. One

morning a group of pigeons flying through hard and so low they had to rise over each fence line, tempted fate by stopping at a puddle in an open paddock for a drink, before continuing on their way home.

Sometimes, at first light, wattlebirds would move in parties from one patch of scrub to another to feed. They would try to pluck up courage by flying out for one hundred metres or so in the direction they hoped to follow and then lose their nerve, swinging back to the safety of the trees. Eventually they would decide it was safe and would fly across nearly a kilometre of open grassland. I could feel the tension mounting in the peregrines above my head as they flexed their muscles and flattened their feathers in preparation for the coming sortie.

Suddenly they would launch themselves, with loud explosions of wings, in quick succession. Like two guided missiles, I watched them home in on their targets. Flying with very fast flickering wing beats, the male led his mate in the chase. Both birds rose slightly above their prey but as they closed on the wattlebirds, the prey turned in a wide arc. The peregrines re-computed their course, taking the 'inside lane'. Swooping under and finally upwards, they both caught one bird each and glided to the nearest fence, landing twenty metres apart, to pluck and eat their prey. These sorties were awesome to watch.

Occasionally, while the peregrines were feeding, wedge-tails would arrive unexpectedly. Once, the bird above my head stopped feeding and I could see it looking skywards, with feathers flattened, while keeping very still. Suddenly the peregrine screamed and flew off. Seconds later the light was dimmed by the eagle's shadow and the whole hide rocked as it hit the top, tearing the prey from its moorings.

With heavy wing beats, the eagle flapped away, carrying the prey in one foot. The female peregrine was joined by her mate in a furious, repeated attack on the giant bird, which dislodged feathers from it. The eagle always won these duels and flew relentlessly away until the defenders gave up and returned to watch and wait for more prey.

A male wedge-tail eagle is an amazingly agile hunter. The main prey of the pair at home consisted of magpies, galahs, crows and ducks. I have watched a male catch mature magpies and ravens with ease, in mid-air. Wedge-tails built up a nest only thirty metres from our house and I used to watch the male fly off through the trees. When departing in a westerly direction he always shot through a very small gap between the branches of two trees. By folding his wings he slipped through at high speed before swooping up into the prevailing wind on the other side. The eagles at home hunted at first light. Well before sunrise, they were often seen flying low cross-country for many kilometres. At the other end of the day they winged to their roost in moonlight, after feeding very late.

Not only do birds like to sit on top of photographic hides but many other animals have enjoyed the comfort inside the weather-proof structures. Sharing my hide with nesting swallows, willie wagtails and grey thrushes has sometimes been inconvenient. They often use the camera-mounting bracket to build a nest. Barn and boobook owls have roosted inside for extended periods, leaving buckets of evidence of their prey, in the form of regurgitated pellets. When using hides at ground level, I've had echidna, sleepy lizards and tiger snakes all invade the premises while I've been in residence.

While photographing a goshawk, a snake nearly caused me to evacuate the hide but I managed to prod it carefully outside with a flash bracket and watched with some relief as it slithered away to annoy someone else. The hides have not been immune from swarms of bees either. After being pestered for some hours by the 'scout' bees, it's a most unpleasant experience when the whole swarm arrives. At times like this mosquito repellent doesn't seem to be good enough and a calm, deliberate exit is called for. Give me a tiger snake any day. Often observations of wildlife unrelated to the job at hand are made from a hide, and it's possible to observe and photograph other birds at the same time.

Many other birds nest quite close to birds of prey. Often at least twenty other species nest within 20 metres of a tree-nesting peregrine. Some, like mudlarks, willie wagtails, parrots and water birds, build as close as a few metres. They all go about their business as usual except for the odd occasion when young are leaving the nest for the first time or smaller species are watering. This provides a photographer with endless opportunities. Some photographs of parrots, cockatoos and 'dicky-birds' (my term for almost any bird that isn't a bird of prey) have been taken out the 'back window', while working on birds of prey. It's amazing how many times the most interesting things happen when there is no camera within reach. The joy of observing such things is often more than adequate compensation. There are so many ideas to follow up, more than is possible in one lifetime.

OSPREY
Pandion haliaetus

An osprey arrives at the nest with a freshly caught fish and its young fledgling screams and begs for food. A camera with a 20 mm lens, mounted upside down on the edge of an artificial nest platform at Ballast Head loading-dock, on Kangaroo Island, made this photograph possible. The young birds had already started to fly and sometimes fed in a tree on the land. The decision not to set up before the young were on the wing was taken in conjunction with the National Parks and Wildlife Service, to minimise any disturbance. In a fortnight, perched on top of the gantry, I witnessed some wonderful interactions between the ospreys and other birds of prey. Almost every day the neighbouring sea-eagle would provoke repeated aerial battles with the ospreys, as it ventured across the invisible territorial border. The sea-eagle would also lock talons with a wedge-tailed eagle for the same reason. A peregrine falcon chased one of the newly fledged ospreys for some kilometres as it returned from an exploratory flight over American River. The osprey had repeated run-ins with the Pacific gulls if they came too close to the nest.

COMMON NAME	Osprey
SCIENTIFIC NAME	*Pandion haliaetus*
DISTRIBUTION	Coastal Australia, except for VIC and TAS
SIZE	Female 60–66 cm, male 50–55 cm
BREEDING	2–3 eggs, nest is a bulky collection of sticks on cliff faces, tall trees, tall poles
DIET	Fish, occasionally sea snakes
STATUS	Moderately common to uncommon

1

AUSTRALIAN SEA-LIONS
Neophoca cinerea

A young seal pup greets its mother in a rock pool as she returns from the ocean. He followed her up on to the rocks where he was allowed to suckle as she lay drying herself in the sun. The pups play in comparative safety in the rock pools, frolicking with each other for hours on end.

Pearson Island, off the South Australian coast, has a breeding colony of this species and a range of animals, from small black pups to adults, can be found. Sea-lions have an eighteen-month breeding cycle, so not all females have pups at the same time.

COMMON NAME	Australian Sea-lion
SCIENTIFIC NAME	*Neophoca cinerea*
DISTRIBUTION	Bass Strait, off SA, south-western WA
SIZE	Males 185–235 cm, females 155–165 cm
BREEDING	One pup
DIET	Squid, diet mostly unknown
STATUS	Sparse

BULL ANT
Myrmecia pyriformis

This creature is one of the minor hazards of camping in the bush. I can remember being stung twice in the same night by one of these. They have formidable weapons at both ends of their bodies. The large prehistoric nippers at the front are backed up by a quite a painful sting in the tail.

They live in small colonies in the ground, often venturing out at night. In spite of their defences trapdoor spiders, such as the mouse spider, feed on them. Along with many other ant species they are an important measure of the health of the environment and a part of the environmental system.

COMMON NAME	Bull Ant
SCIENTIFIC NAME	*Myrmecia pyriformis*
DISTRIBUTION	Widespread
SIZE	2.5 cm
BREEDING	One queen lays all eggs in colony
DIET	General scavenger
STATUS	Abundant

EASTERN GREY KANGAROO
Macropus giganteus

Two bucks fight in a clearing below the Grampian Mountains in Victoria. Eastern greys are the most gregarious of the kangaroo species.

COMMON NAME	Eastern Grey Kangaroo
SCIENTIFIC NAME	*Macropus giganteus*
DISTRIBUTION	Eastern QLD, NSW, VIC (except western deserts), north-eastern TAS, south-eastern SA
SIZE	Head to tail for males 97–230 cm; females 95–186 cm
BREEDING	1 young
DIET	Grasses
STATUS	Abundant

SHORT-BEAKED ECHIDNA
Tachyglossus aculeatus

'Let's shake on it!' the echidna seems to be saying. Echidnas are quite agile and can stand on their hind legs while investigating the bark on trunks of trees for small insects. Their spines give some protection from attack by dingoes and foxes. They can disappear by digging rapidly into the soil when disturbed. They are cute but not cuddly!

An echidna is one animal that has shared my photographic hide. While photographing ducks on the edge of a swamp I watched with amusement as an echidna approached. It pushed its way under the canvas and looked up as it shuffled around under my feet. Then it waddled out and down to the water's edge where it waded into the water for a drink.

COMMON NAME	Short-beaked Echidna
SCIENTIFIC NAME	*Tachyglossus aculeatus*
DISTRIBUTION	Throughout Australia
SIZE	30–45 cm
BREEDING	1 egg laid, after hatching the young is suckled in the pouch for 3 months
DIET	Ants, termites
STATUS	Common

BARN OWL
Tyto alba

This double exposure was created with 'open flash' using two separate infra-red beams. The flashlights fired twice on the same frame as the owl flew through. Barn owls eat mostly rodents, but I have found the remains of small honeyeaters and bats in their regurgitated pellets. One owl roosted for many months in one of my photographic hides. Another owl laid six eggs on the floor of a hide belonging to a friend. He evicted it, but for the next few nights it continually landed on the camera while trying to re-enter the hide.

The barn owl can multiply rapidly when conditions are good, such as during mouse plagues. Mice, like rabbits, can eat themselves out of house and home, and if the population crashes barn owls can, and do, starve.

COMMON NAME	Barn Owl
SCIENTIFIC NAME	*Tyto alba*
DISTRIBUTION	Throughout Australia
SIZE	Female 35 cm, male 34 cm
BREEDING	3–7 eggs, nests in a deep tree hollow or caves
DIET	Rodents, small marsupials, small birds, lizards, beetles, moths
STATUS	Common

WESTERN GREY KANGAROO
Macropus fuliginosus

A young kangaroo jumps over a fallen log as it heads for shelter from an approaching storm. Western greys are usually seen as a small family group with a doe and a joey at foot, with or without a larger buck. There is probably a joey in the pouch and another embryo waiting for the pouched joey to supersede its older brother or sister at foot. Young bucks sometimes gather in mobs of four to eight after weaning, whereas female joeys might tag along with their mother and another joey for some time.

COMMON NAME	Western Grey Kangaroo
SCIENTIFIC NAME	*Macropus fuliginosus*
DISTRIBUTION	Inland southern QLD, western NSW, western VIC, southern SA, southern WA
SIZE	Head to tail for males 94–222 cm; females 97–174 cm
BREEDING	1 young
DIET	Grasses
STATUS	Abundant

RED KANGAROO
Macropus rufus

This image provides a glimpse of a scene that would have been common all along the Murray–Darling River system hundreds of years ago. This is part of a gathering of seventy kangaroos drinking together in a 100-metre stretch of the Darling. The kangaroos' condition was poor and they came in during the hottest part of the day from midday until late afternoon, desperate for water. Little rain had fallen for months.

COMMON NAME	Red Kangaroo
SCIENTIFIC NAME	*Macropus rufus*
DISTRIBUTION	Inland Australia and on coast where rainfall less than 50 cm a year
SIZE	Head and body length for males 93–140 cm; females 74–110 cm; tail length for males 71–100 cm, females 64–90 cm
BREEDING	1 young
DIET	Grasses, plants
STATUS	Abundant

GLOSSY BLACK-COCKATOO
Calyptorhynchus lathami

A pair of glossy-blacks resting in the shade, during the heat of the day. They had fed on casuarina cones for most of the day before flying to drink at a waterhole. The efforts of the Glossy Black-Cockatoo Foundation on Kangaroo Island have helped bring the island species back from the brink of extinction although the battle probably is not won yet.

Some of the remaining casuarina scrub is being conserved and new planting is going ahead. Possums have been excluded from known nest hollows and artificial hollows have been erected to increase the number of nest sites. Scientists are working with the honey bee industry to reduce the honey bee plagues that engulf many of the cockatoos' nesting hollows.

COMMON NAME	Glossy Black-Cockatoo
SCIENTIFIC NAME	*Calyptorhynchus lathami*
DISTRIBUTION	South-eastern QLD, eastern NSW, eastern to central VIC
SIZE	48 cm
BREEDING	1 egg, nests in large tree hollow, breeds March–August
DIET	Casuarina seeds
STATUS	Uncommon

CRIMSON ROSELLA

Platycercus elegans

The Kangaroo Island rosella, shown here inspecting a nest hollow, is isolated from the mainland species. The colour phases this species unveils during its transition to maturity are every bit as beautiful as the plumage of the adult bird. Kangaroo Island is about as far west as the species is found in southern Australia.

COMMON NAME	Crimson Rosella
SCIENTIFIC NAME	*Platycercus elegans*
DISTRIBUTION	Northern QLD tablelands, south-western QLD, eastern NSW, eastern and southern VIC, TAS, south-eastern SA, Mt Lofty Ranges, SA
SIZE	35 cm
BREEDING	4–8 eggs, nest is a hollow in a tall tree
DIET	Seed, fruit, lerps
STATUS	Common

11

SULPHUR-CRESTED COCKATOO
Cacatua galerita

Three sulphur-crested cockatoos take flight. This species has shown its ability to adapt to city life by being quite common in many capital cities in Australia, where its natural food is found in parks, plantations, rivers and gardens. It seems strange to see them circling among the buildings, telephone wires and power lines of the city. Cockatoos make the control of some introduced bulbous weeds difficult, by digging them up just at the best time to spray and breaking many off below the ground. This makes it impossible to apply chemical to the leaf.

COMMON NAME	Sulphur-crested Cockatoo
SCIENTIFIC NAME	*Cacatua galerita*
DISTRIBUTION	Northern NT, eastern QLD, eastern NSW, VIC, TAS, south-eastern SA, Perth.
SIZE	45–52 cm
BREEDING	2 eggs, nests in hollow limb or tree hole high up in gum tree near water
DIET	Seeds of grasses and herbaceous plants, grains, bulbous roots, berries, nuts leaf buds, insects, insect larvae
STATUS	Common

AUSTRALIAN KESTREL
Falco cenchroides

Three young kestrels at the point of fledging sit waiting for their parents to bring food. At this stage they watch other birds and insects throughout the day, carry on furious bouts of wing-flapping exercise, or preen and doze in the sun. When a parent approaches their demeanour suddenly changes to furious begging, combined with whining calls. Much of the food caught by the male is cached by both he and the female in a variety of hide-outs. Dead mice and lizards are left amongst stems of grass, under logs and next to fence posts, sometimes for many hours, before being brought to the nest. This smallest of the Australian falcons feeds on a wide variety of prey — sometimes even starlings and stubble quail are taken.

COMMON NAME	Australian Kestrel
SCIENTIFIC NAME	*Falco cenchroides*
DISTRIBUTION	Throughout Australia except south-western TAS
SIZE	Female 33–35 cm, male 30–33 cm
BREEDING	3–7 eggs. No nest is built — it uses tree hollows, cave entrances, ledges on city buildings or cliffs, nests of raven and other birds
DIET	Insects, rodents, baby rabbits, ground birds, reptiles
STATUS	Common

PACIFIC BLACK DUCK
Anas superciliosa

Camping in a hide overnight is the best way to approach ducks for photography. The black duck is particularly wary and difficult to approach. It's best to let them come to you. It is possible to photograph the same ducks armed with a few bits of bread on a suburban stream or lake, but in the cities it's hard to find pure black ducks. Many have been 'contaminated' by crossing with English Mallard. The mallard hybrids are easy to recognise by their orange legs and bright plumage.

COMMON NAME	Pacific Black Duck
SCIENTIFIC NAME	*Anas superciliosa*
DISTRIBUTION	Throughout Australia except inland deserts
SIZE	47–60 cm
BREEDING	8–10 eggs, nests in scrapes in the ground, in grass, reeds or in tree holes
DIET	Seeds from aquatic plants, aquatic insects, crustaceans
STATUS	Abundant

KOALA
Phascolarctos cinereus

In good weather a large male like this one will sit high in a tree, but during strong wind and rain will move closer to the main trunk into a more secure and sheltered spot. Koalas are solitary animals by nature and rarely tolerate another male in the same tree. This photo was taken with a fish-eye lens, while suspended from a tower, at very close range without disturbing the koala's composure. The slightest vibration on the tree trunk would have had him scrambling up higher or down further.

COMMON NAME	Koala
SCIENTIFIC NAME	*Phascolarctos cinereus*
DISTRIBUTION	East coast of Australia, small areas in southern SA
SIZE	Male 75–82 cm, female 68–73 cm
BREEDING	1 young a year, suckles in pouch for 7 months
DIET	Leaves of particular eucalypt trees
STATUS	Common in limited areas

19

AUSTRALIAN (MALLEE) RINGNECK

Barnardius zonarius barnardi

This species has survived in cleared farming areas where some of the native vegetation still exists. Roadsides provide a living in the Mallee areas where most other trees and shrubs have been ploughed in. In these areas, nesting hollows are scarce and competition fierce with other parrots and starlings. Indeed, the starling opposite is checking out the exactly the same hollow as this ringneck. I recently observed a female wild hybrid, a cross between this species and an eastern rosella. She had mated with a male ringneck and the pair were house hunting together, when I last sighted them.

COMMON NAME	Australian (Mallee) Ringneck
SCIENTIFIC NAME	*Barnardius zonarius barnardi*
DISTRIBUTION	Inland QLD, NSW, SA
SIZE	35–37 cm
BREEDING	4–5 eggs, nests in hollow or hole in tree trunk
DIET	Seeds, fruit, blossoms, fruit buds, insects and insect larvae
STATUS	Common

20

COMMON STARLING
Sturnus vulgaris

This bird should never have been introduced into Australia. It is a pest in horticulture, although it does help control some insect pests, such as heleothis caterpillar and cockchafer grubs. They breed prolifically and form huge flocks that roam the agricultural areas of southern and eastern Australia. When breeding they appropriate nesting hollows of native birds and animals and can fill even quite large hollows with grass and other nesting material. This can make the hollows unsuitable for future use by other parrot species. The starling is a great mimic and can mimic many birds and animals, and even some nocturnal calls such as those of nightjars and foxes.

COMMON NAME	Common Starling
SCIENTIFIC NAME	*Sturnus vulgaris*
DISTRIBUTION	Central QLD, south-eastern Australia, TAS, to Nullarbor Plain, SA
SIZE	20–22 cm
BREEDING	4–8 eggs, nests in tree hollow, cliff hole, crevice in building or bridge
DIET	Insects, spiders, worms, snails, fruit
STATUS	Common

NEST BOX FOR PYGMY-POSSUM

Artificial nest boxes such as this have been provided successfully for pygmy-possums in scrub reserves. Located no more than a metre from the ground in low heath and mallee, they provide a welcome and safe home for tiny possums. One important design detail is the entrance, too small for most birds. The overall size is too small for a hive of honey bees. Bats occasionally visit these nests and roost for short periods.

WESTERN PYGMY-POSSUM
Cercartetus concinnus

A small box placed in the understorey of a pink gum forest has provided an artificial home for this family of pygmy-possums for fifteen years. The three young and their mother nestle in a bed of gum leaves, which are selected in preference to many other species that are available nearby. These delightful creatures are very common on the farm, inhabiting most patches of scrub where the understorey is intact. Their nests are found under logs and mallee stumps as well as in yakka bushes and other dense undergrowth.

COMMON NAME	Western Pygmy-possum
SCIENTIFIC NAME	*Cercartetus concinnus*
DISTRIBUTION	North-eastern VIC, southern SA, south-western WA
SIZE	Head and body length 7.1–10.6 cm, tail length 7.1–9.6 cm
BREEDING	2 or 3 litters of up to 6 young
DIET	Insects
STATUS	Common in its limited area

AUSTRALIAN SEA-LION
Neophoca cinerea

and

PEARSON ISLAND ROCK-WALLABY
Petrogale lateralis pearsoni

A Pearson Island wallaby is disturbed while drinking at a soak and looks up at a huge sea-lion bull as it moves past. The sea-lion has just 'walked' down from its resting place, high on the salt-bush plateau half a kilometre from the sea where it has spent the last few days. They regarded each other for a few minutes and the sea-lion moved on. The wallaby continued to drink for a few minutes, lapping tiny sips from the extremely poor supply.

COMMON NAME	Australian Sea-lion
SCIENTIFIC NAME	*Neophoca cinerea*
DISTRIBUTION	Bass Strait, off SA, south-western WA
SIZE	Males 185–235 cm, females 155–165 cm
BREEDING	One pup
DIET	Squid, diet mostly unknown
STATUS	Sparse

PEARSON ISLAND, SOUTH AUSTRALIA

This uninhabited island off the Great Australian Bight provides a safe, temporary anchorage for fishing boats sheltering from the sometimes huge seas that can be generated by the Southern Ocean. The clear waters around the island are a constant 17°C. The northern section of the island rises out of the water like a water-ringed Uluru. The island is home to a few mammal species, including sea-lions, rock-wallabies and bush rats. Among the rocky outcrops, sea-eagles have massive old nests, constructed of tonnes of sticks. Beneath the dead paperbarks, a favourite feeding tree, the bones of young rock-wallabies, penguins and leather jackets litter the ground. Ravens visit the trees in search of scraps.

27

Over a thirty-five-year period enormous changes have taken place in the upper south-east of South Australia. When I speak with friends who developed country in the same area, we agree that one of our biggest fears was the rabbit. This led to over clearance of areas where rabbits occurred because they used the scrub for shelter and the pasture for feed. We would have left more native bush had the rabbit not been such a potential problem. I had many years of playing god while clearing native bush and replacing it with pastures.

Immediately after being cleared and sown to pasture, the former scrub looked like wonderful parkland. Sometimes quite large belts of bush were left for shade and the area was very appealing to the human eye. Neat fencing and windmills drawing water from a few metres below the ground added to its appeal. This changed once livestock was introduced. The remaining understorey, shrubs and sedges went first; then the trees, unable to take the high fertility transferred to their root system by hundreds of sheep that camped beneath their canopy, died back. This process will continue into the future — few species of eucalypt can withstand this treatment.

water while continuing to flood their downstream neighbours with their own effluent.

Difficult problems now face property owners due to a combination of factors inland from the Coorong. The rising watertable, combined with surface run-off from reliable winter rain, causes widespread flooding which can promote a water induced 'drought' of winter stockfeed. On the other hand, in some areas the water has provided green feed late into the summer season, which more than makes up the difference in overall carrying capacity. A proper drainage system should have been put into action before the land was cleared. Now, an environmental impact study has cleared the way after years of talk. Most people are reluctant to outlay the funds, although the scheme will help make the land agriculturally sustainable. Apart from agricultural land, damage is being done to existing scrub reserves, which are used as dumping grounds for water.

Where does our native wildlife fit into this development? It's getting tougher for some species every year. Ducks and waterfowl may enjoy the flooded land for a few years, but what of the long-term future? Water at the end of many kilometres of drain can end up with a high salt content.

LAND CLEARING AND DEVELOPMENT

Sheep and cattle were not able to thrive in the 'Ninety Mile Desert', as the area was called, before the discovery that trace elements were essential to healthy pasture and livestock. Copper, zinc, cobalt etc. were used along with superphosphate, with spectacular results. During those years, too much native vegetation was cleared. In some areas, non-wetting sands without their natural cover failed to take up rain and flooding of lower lying flats was increased. The water was drained into downstream neighbours with no controls causing more and more loss of productive land the further it went downstream. The upstream farmers are better off with the water gone, but the further downstream one is the worse it gets. Some properties have protected themselves with banks from incoming

When water containing salt at 5000 parts per million is flooded one metre deep over one hectare of land and evaporated, the residue will consist of 50 tonnes of salt per hectare! The mind boggles at the thought of spreading that much salt on the land. Grazing land is only top-dressed with about 100kg of superphosphate per hectare annually.

Very saline water of 6000–9000 parts per million has been used for some irrigated lucerne seed crops in the Keith district for years. The salt stresses the plant, producing high yields of seed. This practice loads the soil with increasing levels of salt, which finds its way eventually, down into the ground water below. It wasn't until the visual effects showed up on the surface that farmers started rethinking the idea.

GREY FALCON
Falco hypoleucos

A young female grey falcon, probably about nine months old, sits contentedly on a log at the edge of the Strzelecki Desert. This beautiful falcon is handsome at all phases of its life, although it loses its attractive spotted appearance as it matures.

After seeing a female swoop into a big mob of galahs from high above, and mortally wound one, I realised its prey extended beyond the smaller birds I had witnessed it hunting before. The grey falcon is rarely seen. Its numbers may have been reduced by the presence of rabbits in its range.

COMMON NAME	Grey Falcon
SCIENTIFIC NAME	*Falco hypoleucos*
DISTRIBUTION	Inland NT, inland QLD, inland NSW, inland VIC, SA, north-eastern WA
SIZE	Female 41–43, male 33–36 cm
BREEDING	2–4 eggs, uses old nests of hawks or crows, breeds July–November
DIET	Birds, small mammals, reptiles, insects
STATUS	Rare

DINGO

Canis familiaris

Vehicle tracks and livestock trails are preferred by dingoes, if they are available, during their travels. This young dog was seen trotting along the middle of the Strzelecki track and moved off to let the vehicle pass. It returned to the road immediately afterwards to continue its journey. The yellow bitch opposite was photographed beneath a black-breasted buzzard's nest as she scouted for prey remains and pellets of regurgitated food. Some dingoes, like foxes further south, have a regular beat around bird of prey nests during the breeding season, looking for scraps.

COMMON NAME	Dingo
SCIENTIFIC NAME	*Canis familiaris*
DISTRIBUTION	Throughout Australia
SIZE	Head and body length for males 86–98 cm, females 88–89 cm, tail length for males 29–38 cm, females 26–38 cm
BREEDING	3–4 pups
DIET	Mammals (such as Common Wombat, Swamp Wallaby, Eastern Grey Kangaroo, Rabbit), reptiles, rodents, birds
STATUS	Common

EMU

Dromaius novaehollandiae

A newly hatched emu chick, out for its first look at the world. The chicks call quite loudly from the egg from a few days before hatching. Because of this, the male will nearly always sit on the eggs until all have hatched. Wedge-tailed eagles easily snatch chicks from the ground if they lag behind the group. In some years, emu chicks form an important part of the eagle's diet throughout the emus' sometimes extended period of breeding. A successful female emu will often settle more than one male on eggs that she has placed in separate nests, consecutively, in the same season.

COMMON NAME	Emu
SCIENTIFIC NAME	*Dromaius novaehollandiae*
DISTRIBUTION	Throughout Australia except southern coast of QLD and northern coast of NSW, central VIC and south-eastern SA
SIZE	2 m from head to tail
BREEDING	7–11 eggs in a grassy bed
DIET	Leaves, grasses, fruits, flowers, insects, seeds
STATUS	Locally abundant

The male emu is attracted by the loud drumming noise of the female that can be heard for some kilometres on a still night. Once a clutch of eggs is laid (usually six to 12 eggs) the male will take over incubation and look after the chicks for over a year until they are independent. I watched one day as an emu left its nest and chased a wedge-tailed eagle that had landed 100 metres from the brooding bird. On a number of occasions I've seen a male with small chicks charge and chase a motor bike and its rider. The unfortunate riders were usually mustering sheep and hadn't seen the camouflaged emu brooding the chicks on the ground.

PEARSON ISLAND ROCK-WALLABY
Petrogale lateralis pearsoni

Three wallabies taking a closer view of the camera and the person behind it. Their normal feeding behaviour is difficult to photograph due to their curiosity. The salt bush plateau caters for some of their food. The sea provides an unusual background compared with that of their inland relations. Living on a windswept island in the Great Australian Bight provides special problems for some individuals. The salt spray, driven by the wind, causes matting of the fur of some animals living on the western slopes and the salt prevents these animals from grooming their coats properly. Their drinking water is already quite saline so they may not be able to tolerate much more.

The wallabies on these islands have been isolated from the mainland for some 25,000 years. Quite simple but drastic biological control seems to prevent the species from over-populating the islands. The supply of fresh water is limited in summer and the competition for this is won by the most dominant and fittest animals. The old and infirm and many of the young perish each year and this allows the native vegetation to maintain a balance with the population.

COMMON NAME	Pearson Island Rock-Wallaby
SCIENTIFIC NAME	*Petrogale lateralis pearsoni*
DISTRIBUTION	Pearson Island, SA
SIZE	Head and body 45–53 cm, tail 48–60 cm
BREEDING	1 young
DIET	Grasses
STATUS	Rare

NEW ZEALAND FUR-SEAL
Arctocephalus forsteri doriferus

Limpid eyes look out from a hiding place in the rocks, as a group of younf fur-seal pups hide. Their 'captors' were CSIRO scientists, who were tagging the young seals for a population survey.

Life can be tough for these creatures. I watched a lone young fur-seal riding a wave towards the rocks in a huge sea. As it approached a cliff, the wave it rode hit the backwash from the rocks and it was flung high in the air. The exposed coast where they haul out to the land would be a frightening prospect for any human swimmer. An inquisitive pup came

COMMON NAME	New Zealand Fur-seal
SCIENTIFIC NAME	*Arctocephalus forsteri doriferus*
DISTRIBUTION	Southern central SA, southern WA
SIZE	Male 150–250 cm, female 130–150 cm
BREEDING	1 pup is born end of November–December, mating occurs about eight days later. Males leave colony in January, pups are weaned by August.
DIET	Squid, octopus, barracouta, other fish, lampreys, lobsters, penguins
STATUS	Sparse

within a few millimetres of the camera as I lay on the rocks at Cape Du Couedic. His breath fogged the wide-angle lens when he exhaled loudly at the sound of the shutter firing. Their slick appearance in the water changes to a fluffy and furry look as they dry out on land.

The Casuarina Islands at the south-western corner of Kangaroo Island provide a platform for a breeding colony for the fur-seals. This species is on the way back after many years of persecution at the hands of hunters.

WHITE-FACED HERON
Egretta novaehollandiae

A heron stands wounded beneath a power line, the scourge of flying wildlife. Single wire, earth return lines are one of the major causes of death by collision of birds in Australia. Eagles, falcons, ducks, pigeons and many other birds find the wires difficult to observe at certain times and in certain light. Even the birds of the inland are being subjected to the same treatment, with the power lines servicing much of the outback New South Wales and power stations being built in the Cooper gas fields.

COMMON NAME	White-faced Heron
SCIENTIFIC NAME	*Egretta novaehollandiae*
DISTRIBUTION	Throughout Australia except central WA
SIZE	60–70 cm
BREEDING	Up to 7 eggs, nest is a loose platform of sticks in a tree
DIET	Crustaceans, squid, fish, insects, amphibians, spiders, snails, worms
STATUS	Common

BILBY
Macrotis lagotis

The bilby's decline in Australia is testament to man's direct interference by the introduction of feral animals. Within ten years of the rabbit arriving in any given region the bilby disappeared. Rabbits competed directly with it for food. The introduced fox and cat have eaten the stragglers. The bilby is a truly threatened species.

COMMON NAME	Bilby
SCIENTIFIC NAME	*Macrotis lagotis*
DISTRIBUTION	Inland north-western WA, inland NT, south-eastern QLD
SIZE	Head and body length for males 30–55 cm, females 29–39 cm; tail length for males 20–29 cm, females 20–28 cm
BREEDING	2 young which leave pouch after 75 days
DIET	Insects, insect larvae, seeds, bulbs, fruit, fungi
STATUS	Rare

STRIATED PARDALOTE
Pardalotus striatus

Below, a pardalote looking into its nest hollow. I recorded its call and when playing the sound back, the male landed on my hand. When it saw its reflection in the camera lens, a split second later, it turned on a display, showing the red spots on its wings. While I sit high in the treetops these little birds make good company. They move through the branches of the pink gums feeding on lerps, which can devastate the trees in some seasons.

COMMON NAME	Striated Pardalote
SCIENTIFIC NAME	*Pardalotus striatus*
DISTRIBUTION	Throughout Australia, except a strip in central Western Australia extending to the coast
SIZE	9–11.5 cm
BREEDING	3–5 eggs, nests in large colonies in hollows in trees and in earth banks
DIET	Insects in tree foliage
STATUS	Common

BANDED STILT
Cladorhynchus leucocephalus

Birds from part of a huge flock take to the air. After spending most of the morning on a hot summer's day swimming in saline water while foraging for brine shrimps the birds gather on banks to preen and rest. Small parties fly in, low to the water, from the surrounding salt swamps to join the main gathering. Their form is so different in flight from on water or land. Their long legs balance the neck, head and long beak. With wings swept back in an arc, they race across the water.

Peregrine falcons hurl themselves at a flock, which can form an extremely dense group. There always seems to be a straggler, which is picked off from behind and carried off to a lonely bank. The peregrine will often just eat the head and neck and hide the rest next to a bush, or rock, for half a day out of sight of the swamp harriers before returning to finish it off.

COMMON NAME	Banded Stilt
SCIENTIFIC NAME	*Cladorhynchus leucocephalus*
DISTRIBUTION	Western NSW and VIC, eastern SA, south-western and inland WA
SIZE	36–45 cm
BREEDING	3–4 eggs in a small depression in the ground
DIET	Brine shrimp, small crustaceans
STATUS	Locally abundant to uncommon

BARN OWL
Tyto alba

A pair of corellas shared their nest hollow with this owl during the day and when the owl was disturbed, it flew into a tree where it stood gazing through the foliage. A brown falcon nesting nearby interrupted its gaze and chased the owl until it disappeared into another nest hollow.

COMMON NAME	Barn Owl
SCIENTIFIC NAME	*Tyto alba*
DISTRIBUTION	Throughout Australia
SIZE	Female 35 cm, male 34 cm
BREEDING	3–7 eggs, nests in a deep tree hollow or caves
DIET	Rodents, small marsupials, small birds, lizards, beetles, moths
STATUS	Common

TAMMAR (DAMA) WALLABY
Macropus eugenii

This species has almost disappeared from Australia's southern coastal areas, although it is abundant on Kangaroo Island and some offshore islands.

The wallabies are preyed on by wedge-tails and sea-eagles. Luckily, there are no foxes on the island to attack their young in the dense cover, where they are safe from aerial attack. In some areas however, without culling by farmers the remnants of native vegetation would be devoid of many edible species due to over-grazing. Like the larger species on the mainland, inside the dog fence, they are thriving due to the comparatively good balance of scrub, pasture and artificial water supply.

COMMON NAME	Tammar (Dama) Wallaby
SCIENTIFIC NAME	*Macropus eugenii*
DISTRIBUTION	Eyre Peninsula and offshore islands (including Kangaroo Island) in SA, south-eastern WA and offshore islands
SIZE	Males 59–68 cm body length, 38–45 cm tail length; females 52–63 cm body length, 33–44 cm tail length
BREEDING	1 young
DIET	Grasses, herbs, leaves, fruit
STATUS	Common in its limited range

RED-RUMPED PARROT
Psephotus haematonotus

A female looks for landing space on a rock in a waterhole in the Strzelecki Creek, in northern South Australia. Feeding on grassland in the open and being direct fliers puts these parrots on the menu of most species of falcons. Near our farm male peregrines and Australian hobbies caught mostly starlings, but red-rumps came next on their list of preferred prey while feeding brooding mates. If the peregrines or hobbies are attacking a flock of parrots feeding on the ground they skim the grass tops at high speed, with their wings folded back for some distance as they approach, relying on a surprise attack. If the stoop is unsuccessful at first, they have the agility to turn quickly and pursue their prey again.

COMMON NAME	Red-rumped Parrot
SCIENTIFIC NAME	*Psephotus haematonotus*
DISTRIBUTION	Southern QLD, inland NSW and VIC, eastern SA
SIZE	27 cm
BREEDING	4–5 eggs, nest in hollow limb or hole in a tree
DIET	Grain, seeds, shoots, leaves, flowers
STATUS	Common

THE FERAL (EUROPEAN) RABBIT
Oryctolagus cuniculus

My heart sinks when I think of the damage done by this introduced animal, particularly in the delicately balanced semi-desert regions of Australia. It is amazing to see small, fenced experimental plots in the Flinders Ranges and north of Alice Springs, where rabbits and livestock have been excluded from the surrounding country. The difference is staggering, both in variation and health of native plant species.

COMMON NAME	Rabbit
SCIENTIFIC NAME	*Oryctolagus cuniculus*
DISTRIBUTION	Throughout Australia except north-eastern WA, northern NT, northern QLD
SIZE	Male 35–42 cm, female 35–41 cm
BREEDING	4–5 kittens per litter, up to 5 litters a year
DIET	Green grass and herbage
STATUS	Abundant

WILLIE WAGTAIL
Rhipidura leucophrys

Two photographs, taken a split second apart, show how quickly a willie wagtail can move. One second it's sitting on the falcon's shoulder and the next it's on the branch beside. The falcon is too slow to grab it. The brown falcon, which doesn't build its own nest, appropriated a huge old wedge-tail's nest and the willie wagtail built its cup-shaped nest, of spiders web and fine stems, in the stick material under the rim. Willie wagtails often nest close to bird of prey nests and harass the main tenant constantly. It is a very courageous little bird when its family is at stake.

COMMON NAME	Willie Wagtail
SCIENTIFIC NAME	*Rhipidura leucophrys*
DISTRIBUTION	Throughout Australia, except southern TAS
SIZE	19–21.5 cm
BREEDING	2–4 eggs, builds nest on tree branch
DIET	Insects
STATUS	Abundant to common

BROWN FALCON
Falco berigora

The brown falcon is a unique bird that hunts a wide range of prey including reptiles, small mammals, birds and insects. It is the most fleet of foot of the falcons and often strides around in long grass hunting for prey. Much of their hunting is initiated from a perch, waiting for movement to betray the presence of prey. The falcon's head weaves and bobs up and down as it uses its monocular vision to pinpoint an animal before flying after it. These falcons can hold their position in mid-air and will often hang by their wings in a strong wind, like a kite on a string. They rely on birds such as ravens, crows and magpies to provide nests, although wind can often demolish these structures before their brood is fledged.

COMMON NAME	Brown Falcon
SCIENTIFIC NAME	*Falco berigora*
DISTRIBUTION	Throughout Australia
SIZE	Female 48–51 cm, male 41–45 cm
BREEDING	2–5 eggs, usually in old nest of another hawk species, sometimes in open tree hollow or builds its own large stick nest
DIET	Small mammals, small birds, lizards, snakes, caterpillars, grasshoppers, crickets, beetles
STATUS	Common

BLACK FALCON
Falco subniger

I sat on the ground with my back to a power pole watching with field glasses and noticed an almost invisible cross in the sky that marked the position of a black falcon. Far below, perched on a single wire that slashed across the countryside, a brown falcon watched a stubble quail on the ground. For a while, the quail was invisible, crouching motionless. Then it moved and the brownie launched itself off the wire, and with a controlled, gliding stoop landed on its prey.

Immediately the black dropped like an arrow, in a vertical dive. The brownie, having broken the quail's neck with one bite of its specially designed beak, started to run, with the prey in her mouth, towards a good hiding place. She would cache the food for later in the day.

The black was in full pirating mode and sped down towards her victim. Suddenly there was a loud twang, which vibrated down the power line. I could feel a shudder through my back. A few dark feathers hung in the air as the wounded falcon spun out of control to earth, as if shot. The brownie took off with its prey in one foot and flew steadfastly away as I went to investigate.

The black falcon had broken its wing. Power lines kill hundreds of birds in this way. I have found pelicans, swans, eagles, herons, falcons, magpies and racing pigeons dead or wounded beneath the lines.

COMMON NAME	Black Falcon
SCIENTIFIC NAME	*Falco subniger*
DISTRIBUTION	Interior of NT, coast and interior of QLD (except east coast of Cape York Peninsula), interior of NSW, coast and interior of VIC, SA, south-western and north-eastern WA.
SIZE	Female 52–56 cm, male 45–54 cm
BREEDING	2–4 eggs, uses old nests of other species of hawk, raven or crow
DIET	Birds, rabbits, rats
STATUS	Rare

GALAH
Eolophus roseicapilla

An immature galah alerts other approaching birds to its presence with open-mouthed threat, warding off an unwanted landing. This is an example of the value of manual exposure when photographing birds in flight with ever-changing backgrounds. As birds fly from bright sky into shadow, 'automatic' exposure can let the photographer down.

COMMON NAME	Galah
SCIENTIFIC NAME	*Eolophus roseicapilla*
DISTRIBUTION	Throughout Australia, except Cape York Peninsula, central Australia and inland from the Great Australian Bight
SIZE	35–36 cm
BREEDING	2–6 eggs, nests in hollow limb or hole in a tree
DIET	Fallen seeds
STATUS	Common

PEREGRINE FALCON
Falco peregrinus

A falcon plucks a dead pigeon from shallow water after knocking it from the air, and carries it away to shore. She will decapitate it, eating the head and neck, then pluck the flight and tail-feathers and gut it before flying off. Holding the prey firmly in her talons, the falcon will tuck the dead bird under her tail and, after initially quite laboured flight, she will circle easily into thermals above the open country. She will gain height with amazing rapidity and grace until she is just a dot in the sky. Then, after seeing the way is clear of eagles and black falcons, she will glide at high speed across the sky to her nesting territory some kilometres away. She will hide the food before taking it eventually to where her young are close to fledging.

COMMON NAME	Peregrine Falcon
SCIENTIFIC NAME	*Falco peregrinus*
DISTRIBUTION	Throughout Australia
SIZE	Female 40–50 cm, male 35–42 cm
BREEDING	2–4 eggs, use the same nest site (recess in cliff, hollow in large tree, abandoned large nests of other birds) year after year
DIET	Small to medium-sized birds, rabbits, other small mammals
STATUS	Moderately common

AUSTRALIAN MAGPIE
Gymnorhina tibicen

A softly, softly approach was taken in getting close to this magpie for a photo, and quite quickly the family was sitting on top of the hide, preening and singing in between feeding forays.

The magpie can be a fearsome predator at times, relentlessly pursuing small birds. Amazingly, they sometimes kill and eat birds such as starlings and stubble quail as well as smaller pipit-sized birds. Magpies are well known for their attacks on intruding humans when nesting. I can remember as a child wearing a helmet like 'Ned Kelly's', made out of a tin bucket with eye holes cut in it, which allowed me to negotiate the parklands without getting pecked on the head.

Magpies develop quite a set routine when rearing their young. After delivering food to the young the parent would take the faecal sac and carry it away from the nest. A zig-zagging, aggressive flight would ensue as the magpie chased away the little ravens that had strayed too close to the nest. The bird would then start searching for food.

Starting hundreds of metres away it would walk back to within a few metres of the nest, collecting mostly small wolf spiders and lining them up in its bill. It would fly the last few metres to the nest and reveal up to thirty spiders, grubs which it would share out among the young. Then away it would go, with a mad dash to regain its territory before hunting again.

COMMON NAME	Australian Magpie
SCIENTIFIC NAME	*Gymnorhina tibicen*
DISTRIBUTION	Throughout Australia except WA to the coast, northern NT, Cape York Peninsula
SIZE	36–44 cm
BREEDING	3–5 eggs, in a nest of sticks and plant stems in a tall tree
DIET	Insects, worms
STATUS	Locally abundant to common

WOLF SPIDER
Lycosa sp.

This little wolf spider, or claypan spider as I sometimes call it, lives in very hard, flat claypan surfaces and sits above its burrow on dull days and at night waiting to intercept insects. During the day, it shuts the trapdoor on top of the burrow to maintain enough humidity for survival in the sometimes extreme conditions. The camouflage of the entrance is near perfect, with the hinged door fitting snugly. The burrow of this species is lined with strong silk. Some species deliberately position a small stone on the top of the lid, which makes the burrow even harder to find.

DRIFTWOOD IN THE GREAT SANDY DESERT

A dead tree on the floor of a claypan in the Great Sandy Desert. The night before I took the photo this claypan was a natural airstrip. After two inches of torrential rain, it turned to mud and only a narrow strip remains dry, around the perimeter. I wonder what happened to the wolf spiders, whose eyes shone in the torch light before the rain, while sitting at the entrance to their burrows, ready to pounce on insects.

PEREGRINE FALCON
Falco peregrinus

In early winter, a female peregrine looks close to perfection, with most of its feathers intact. It is difficult to get the under-wing detail in a photograph of a wild bird of prey because you are relying on reflected light from the ground. The wind direction and strength proved the most important factors in obtaining this photograph. The magnificent peregrine falcon is undoubtedly my favourite bird.

COMMON NAME	Peregrine Falcon
SCIENTIFIC NAME	*Falco peregrinus*
DISTRIBUTION	Throughout Australia
SIZE	Female 40–50 cm, male 35–42 cm
BREEDING	2–4 eggs, use the same nest site (recess in cliff, hollow in large tree, abandoned large nests of other birds) year after year
DIET	Small to medium-sized birds, rabbits, other small mammals
STATUS	Moderately common

One of my old friends keeps saying to me 'we have poisoned all the wildlife. I can no longer hear the bats as they fly around the paddocks at night'. In fact, his hearing has failed (all those years driving a noisy tractor without ear-muffs). Bats are still around, and may actually be more useful than birds for insect control in large treeless paddocks. But bats could be threatened if they do not have enough places to roost. Most species of bats use hollow trees for roosting. These are disappearing rapidly, with thousands of dead trees removed for firewood every year. The removal of dead timber on the ground also eliminates hiding places for small reptiles, including frogs and lizards.

The remaining hollows are under threat from a 'friend of the farmer', the introduced European honey bee. In areas where specialist crops such as lucerne seed are produced, aerial spraying of insect pests is carried out at night to avoid killing the 'friendly' bees that are pollinating the lucerne. At the same time of night, bats are out catching the pests being sprayed. This creates an interesting quandary.

unintentional poisoning can occur through misuse of farm chemicals. Proper disposal of contaminated sheepskin and tails removed from lambs at lamb marking and mulesing is important. This material should not be left for ravens and magpies to 'clean up'.

Thousands of acres were once sprayed with DDT in agricultural areas every year to control red-legged earth-mite, bud worm and other pests. It has been proved to cause great harm to animals and birds high up the food chain, while at the same time the insect pests quickly turned out new generations that were immune to its effect.

Fox trapping, using rabbit traps set around dead rabbit or kangaroo, leads to the unintentional trapping and injury of many birds of prey. The practice is probably only resorted to when fox skins are valuable on the world market.

Thirty per cent of the land on our property was set aside as native bush and fenced to keep rabbits and livestock out. Such areas of native bush are not compatible with livestock, or even overstocking of native animals.

ON THE FARM

Many farmers are now employing crop monitors to help in the decision making process regarding spraying. Luckily the soft approach that is being taken with reductions in chemical use, due to monitoring, is not only economically sound but is proving to be good for the environment. In lucerne seed crops, lower strength as well as two-thirds less chemical is being used than a few years before. Lighter chemical use has resulted in higher survival rate of the natural predators and parasites of pests.

Shelter belts and reserves of scrub help retain birds and bats that help control pests. Ravens eat bud worm and other pests and are being recognised as important indicators of insect pests in crops. Ravens do more good than harm and should be protected. In many areas they provide nest sites for most of the falcon species.

It is sad to hear reports of farmers who still try to poison ravens and cockatoos. Poison can affect non-target animals such as birds of prey and

Once areas are isolated, in the form of islands, the diversity of plant and animal life will be eroded gradually and many species of both plant and animal will die out. Surplus kangaroos need to be culled so they don't damage the scrub areas. Weed invasion from the perimeter is difficult to control. The spraying of weeds in scrub can kill the very plants one is trying to protect.

Connecting corridors of bush between the islands of scrub were not left in the original clearings because of the threat of rabbits, but later, with good control of rabbits, corridors have been planted with seed, and many native species have been successful. Birds and animals make good use of the corridors. The plantings have had mixed results depending on the season and weed control. Tree planting and fencing of scrub areas is costly but the connection of isolated areas of scrub is vital to the diversity and health of the animal populations.

SOUTHERN BELL FROG
Litoria raniformis

Drainage, introduced fish species and chemical use may be having an effect on the population of this species of frog in agricultural areas of southern Australia. When I was young, it was always present under the float covers of stock-watering troughs, south of Naracoorte in South Australia. It is one of the most beautiful 'tree' frogs but is in fact more a ground dweller and would probably find it easier to climb into a water trough than up a tree. The frogs' chorus in the watercourses inland from the Coorong can be deafening in early spring. It's a truly amazing sound. Even this population is threatened by increased salt from deep drainage channels and the use of herbicides.

COMMON NAME	Southern Bell Frog
SCIENTIFIC NAME	*Litoria raniformis*
DISTRIBUTION	South-west NSW, VIC, TAS, south east SA
SIZE	8 cm
DIET	Insects, spiders, larvae
STATUS	Abundant

SHINGLE-BACK LIZARD
Tiliqua rugosus

My tent remained in the bush for a month and when it was moved, I found a lizard had made camp beneath it. When it was unearthed, its colour blended well with the surrounding sedges. The banded designs of the Olary Ranges' specimens change to solid copper-brown in the sand hills near Menindee. At home on our farm, the lizards were always dark, greenish to black in colour. It is interesting to watch them grazing on succulent plants and native berries. They enjoy eating the introduced Cape Weed, tearing off the daisy-like flowers and swallowing them whole. Around farm houses these lizards become almost domesticated, lining up at the 'dinner table' with the dogs to share their food.

COMMON NAME	Shingle-back Lizard
SCIENTIFIC NAME	*Tiliqua rugosus*
DISTRIBUTION	Central and southern QLD, western VIC, southern SA, southern WA
SIZE	25 cm
BREEDING	1–3 young
DIET	Insects, arthropods, snails, carrion, flowers, fruit, berries
STATUS	Abundant to common

BEARDED DRAGON
Pogona vitticeps

A bearded dragon, at home in the Strzelecki Desert. The black-breasted buzzard and wedge-tailed eagle find these easy to catch, once they are sighted. Although well camouflaged they give themselves away by moving around. I watched a wedge-tail stoop in from some kilometres away to pirate a freshly caught dragon from a whistling kite, which had no defence against such a powerful bird. The air tore loudly through the eagle's plumage as it rocketed past, just metres above my head. My first instinct was to dive for cover, but it was wonderful to see the grace and control of the eagle at such close range.

COMMON NAME	Bearded Dragon
SCIENTIFIC NAME	*Pogona vitticeps*
DISTRIBUTION	Inland QLD, inland NSW, north-western VIC, north-east SA, south-east NT
SIZE	20 cm
BREEDING	1 or 2 clutches of 11–25 eggs
DIET	Insects
STATUS	Common

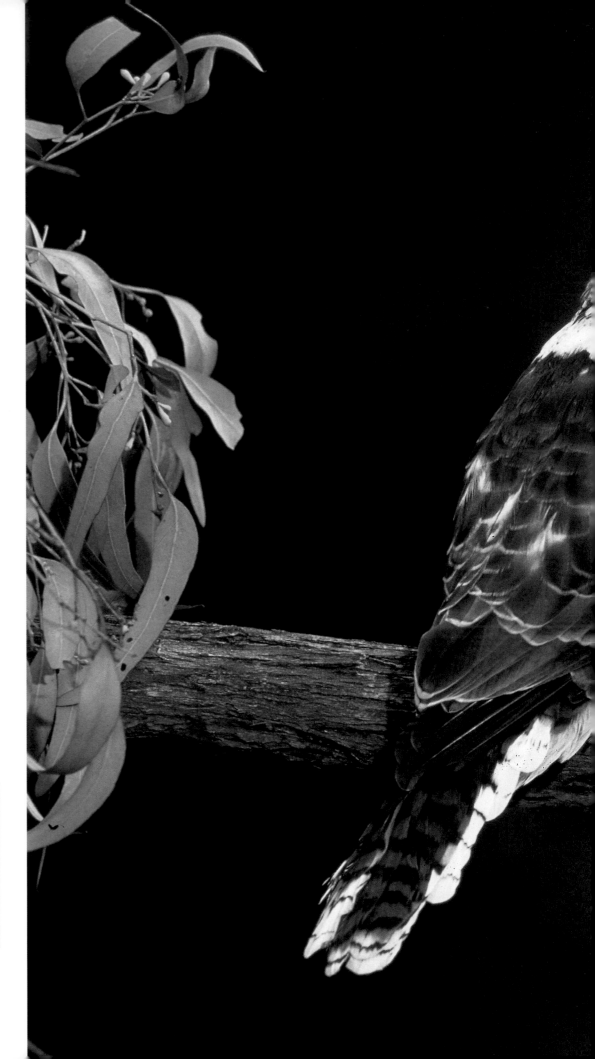

LAUGHING KOOKABURRA
Dacelo novaeguineae

A pair of kookaburras 'sing' to advertise their presence near nest holes and favourite perches throughout their territory. The sound of other family groups nearby sends them into a frenzy. The largest of the kingfisher family, this bird is a fearsome predator of small birds, reptiles, mammals and insects. A 60-litre drum, positioned in a tree, was adopted by one pair whose nest had been commandeered by bees the year before. It created an opportunity to photograph, and see close up, the centipedes, beetles, yabbies and small lizards that formed the prey items brought to a nest.

COMMON NAME	Laughing Kookaburra
SCIENTIFIC NAME	*Dacelo novaeguineae*
DISTRIBUTION	Eastern QLD, eastern NSW, VIC, TAS, south-eastern SA, south-western WA
SIZE	45 cm
BREEDING	1–4 eggs, nests in large cavity in tree trunk or branch, termite mound
DIET	Insects and other invertebrates, snakes, lizards, small birds, rodents
STATUS	Common

BLUE GUMS,
KEITH, SOUTH AUSTRALIA

The blue gum grows in quite shallow clay soils where water gathers during winter. Many of this species are blown over by strong spring winds while the roots are inundated. Very few hollows appear in these trees but they provide food in the way of blossom and fruit for lorikeets, parrots, woodswallows and honeyeaters, and insects for bats. They are more resistant to the increases in fertility due to fertiliser use and livestock damage than some other gums and survive long after the stringy bark and pink gums have died in pastured paddocks. The understorey plants that were left surrounding each tree have been eaten by stock and the water lies on an area that once was covered with melaleuca bushes. The wet areas need 'controlled' drainage or the remaining topsoil will blow away in summer.

LESSER LONG-EARED BAT
Nyctophylus geoffroyi

One evening a storm was brewing in the western sky and the air was alive with insects as I sat next to a small campfire on the edge of the Strzelecki Desert. Insects were attracted to the light of the fire and after a lap or two most of them perished in the flames. With a flurry bats repeatedly homed in and snatched the insects from quite close to the flames.

A few years later in the shearing shed one night I noticed bats flying around inside. I remembered the incident up north and used a candle to light the building for a photo. Moths were drawn to the candle and the bats pursued them. The camera and flash were triggered by an infra-red beam, and the shutter remained open for two seconds to record the candle flame.

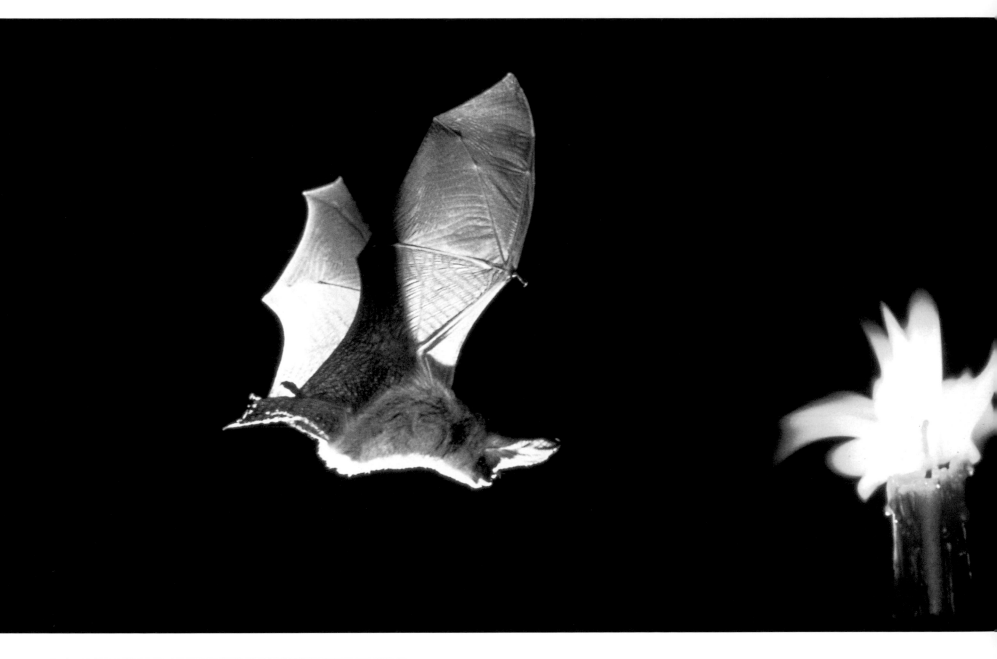

COMMON NAME	Lesser Long-eared Bat
SCIENTIFIC NAME	*Nyctophylus geoffroyi*
DISTRIBUTION	Throughout Australia except coast of QLD and Cape York Peninsula
SIZE	Head and body length 4–5 cm, forearm length 3.3–3.9 cm
BREEDING	Maternity colonies of 10–100 young
DIET	Insects
STATUS	Abundant

A bat flies through the door of the shearing shed carrying a dung beetle. It would then circle the inside of the shed dropping some of the inedible parts of the beetle on the floor as it prepared its meal. To do this it gripped the beetle using the hind feet and the skin between the hind legs, while it tore off the wing cases with its teeth. Often the bats would bring beetles and large moths inside and would hang from the rafters while they fed. This was observed using candlelight, which the bats tolerated, unlike torch light that drove them to darker corners. Often the floor of the shed was covered with 'spare parts' of moths and beetles.

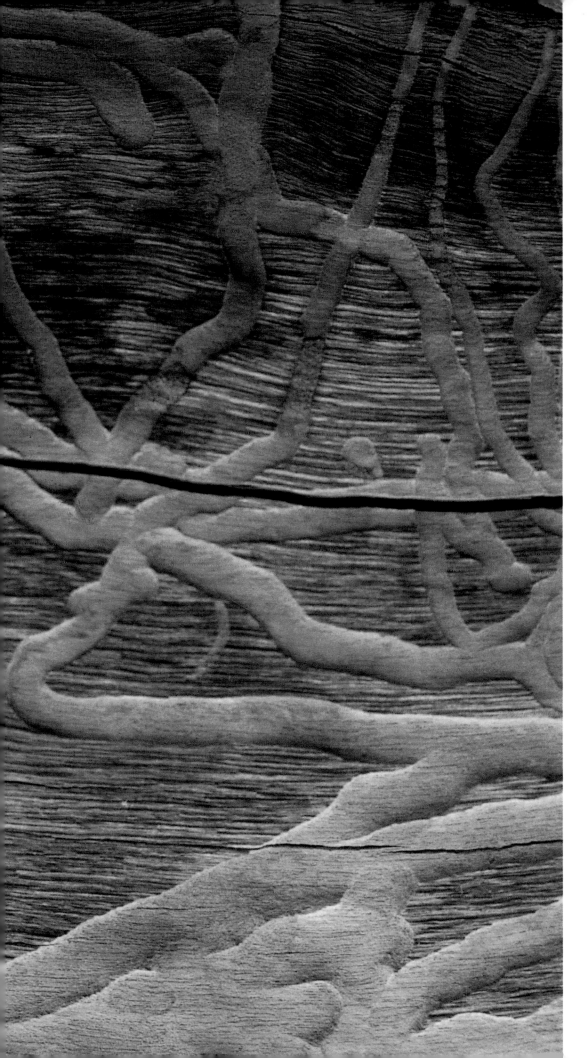

GRUB ART

The trunk on a fallen pink gum is inscribed with a unique design by borers that have mined beneath the bark in search of food.

This was a large, healthy tree twenty years ago, in a reserve that had never been grazed with livestock. In spite of this the tree died from a combination of factors. The sand hills upstream were recently covered with lush lucerne pastures and a sustainable system seemed to have been achieved. Disaster struck in the 1970s, when the alphalfa aphid became a problem. The lucerne died and the increased runoff flooded the tree roots for too long. Stress weakened it, leaving it open to attack by longicorn beetle larvae. The tree, which provided hollows for many creatures, lost its grip and fell. This sad story is to be repeated millions of times.

LETTER-WINGED KITE
Elanus scriptus

The Letter-winged Kite, a rarely seen nocturnal and twilight hunter whose fortunes rise and fall largely on the population of the long-haired rat, *R. villosissimus*. This species has the ability to breed repeatedly in colonies when prey is plentiful in good seasons. Like the barn owl, its numbers dwindle and the birds will disperse when the rats decline. The photograph was taken on the edge of Sturt's Stony Desert in South Australia. Although nocturnal hunters, these birds defend their nest throughout the day from other raptors. The more common Black-shouldered Kite, a close relative, is often seen hunting on dull days along roadsides in many parts of Australia because its favoured prey, mice, come out to feed in these conditions.

COMMON NAME	Letter-winged Kite
SCIENTIFIC NAME	*Elanus scriptus*
DISTRIBUTION	Inland eastern NT, western QLD, north-western NSW, north-eastern SA
SIZE	Female 35–38 cm, male 33–36 cm
BREEDING	3–6 eggs, nest of twigs
DIET	Rodents, reptiles, insects
STATUS	Locally abundant to uncommon

LIGHTNING

Thunderstorms can give and take life depending on their mood. Dry storms often result in fires. Many fires start and then are extinguished by rain almost immediately, leaving only a small burnt-out area. Whether to allow fire in nature reserves is a difficult question, as some trees and plants will benefit and others will not. In small isolated reserves a fire can be disastrous for the animal life as some species may be eliminated, with no way of returning if similar vegetation does not exist nearby. In a case such as this, roadside vegetation becomes an important resource.

KOALA
Phascolarctos cinereus

A young koala clings to its mother's back 15 metres above the ground. At this stage of its life, the young move away from their mothers quite often to search for their own supply of gum leaves. This family had been in the same tree for almost three months and would continue to graze until they had killed the tree. Five big dead manna gums stand nearby in silent memorial to human interference. The koala was introduced to Kangaroo Island in the 1920s and has caused widespread damage to the native vegetation by over-grazing its favourite species of gum.

COMMON NAME	Koala
SCIENTIFIC NAME	*Phascolarctos cinereus*
DISTRIBUTION	East coast of Australia, small areas in southern SA
SIZE	Male 75–82 cm, female 68–73 cm
BREEDING	1 young a year, suckles in pouch for 7 months
DIET	Leaves of particular eucalypt trees
STATUS	Common in limited areas

BLUE-WINGED KOOKABURRA
Dacelo leachii

The first few times I heard this kookaburra call I thought it was a brolga. The raucous call, sounding like a composition from a computerised synthesiser, just didn't seem to fit a bird of this size.

As an experiment I've played the calls of the Laughing Kookaburra to this species and prompted an immediate response — loud and full song. Conversely the Laughing Kookaburra paid little attention to the recordings of the Blue-winged.

COMMON NAME	Blue-winged Kookaburra
SCIENTIFIC NAME	*Dacelo leachii*
DISTRIBUTION	North-western WA, north-eastern WA, northern NT, western QLD, eastern QLD
SIZE	40–45 cm
BREEDING	1–4 eggs, nests in hollow tree cavity, termite mound or baobab tree hollow
DIET	Snakes, lizards, rodents, small birds
STATUS	Common

BROWN GOSHAWK
Accipiter fasciatus

A young bird, only recently independent, chases almost anything that moves. During this phase it will chase parents, and other birds and animals, both smaller and much larger than itself. I once watched a young goshawk land on the back of a fully grown grey kangaroo and ride it for 30 metres, prompting the surprised animal to accelerate away to cover.

Young birds have very distinct, bold markings, which give way to mature plumage after two years. This bird can upset aviary birds by just appearing in the sky above. When a hawk actually lands on the cage and harasses the occupants, the 'justice' metered out by the owner depends to some extent on the value of the birds being harassed. As the goshawk is a protected species, the best solution is to take preventative measures like shade-cloth and roofing to protect aviary birds.

The use of cut-outs and kites, emulating this bird, built to scare small pest birds in vineyards and orchards is useless. These birds can tell the difference between a hunting bird and a well fed one, let alone the difference between a bit of plastic and the real thing.

COMMON NAME	Brown Goshawk
SCIENTIFIC NAME	*Accipiter fasciatus*
DISTRIBUTION	Throughout Australia except some inland deserts
SIZE	Female 45–55 cm, male 38–45 cm
BREEDING	1–5 eggs, nest of sticks and twigs in tree fork
DIET	Birds, small mammals, reptiles, frogs, large insects
STATUS	Common

LYREBIRD
Menura novaehollandiae

Lyrebirds can develop an amazing repertoire of mimicked calls. In the wild, lyrebirds even have the ability to mimic machine-made noises like chain saws and trucks, as well as many other bird and animal sounds. Their mating display is one of the most vibrant events to be seen and heard in the bird world.

COMMON NAME	Lyrebird
SCIENTIFIC NAME	*Menura novaehollandiae*
DISTRIBUTION	East coast of southern QLD, east coast of NSW, east coast of VIC to Dandenong Ranges, western TAS
SIZE	Male 80–90 cm, female 74–84 cm
BREEDING	1 egg, in bulky nest of sticks, ferns and mosses on the ground or on tree-stumps
DIET	Insects, worms and other invertebrates
STATUS	Moderately common

BRUSH-TAIL POSSUM
Trichosurus vulpecula

A possum watches a more dominant animal on the ground below, from the safety of a tree. The competition for territory among possums is fascinating to watch. Just a stare from a dominant animal is often enough to put a lesser mortal to flight. Possums understand each other's body language from a surprising distance and when ignored it can lead to a fight. Many pet possums are collected from road killed females and according to recent surveys most are killed or die when they are released to the wild. This territorial behaviour is common to most of the mammals and birds I've watched.

COMMON NAME	Brush-tail Possum
SCIENTIFIC NAME	*Trichosurus vulpecula*
DISTRIBUTION	Eastern QLD (except Cape York Peninsula), eastern NSW, VIC, TAS, central Australia, south-western WA
SIZE	Head and body length 35–55 cm, tail length 24–40 cm
BREEDING	A single young is born every year; it spends 4–5 months in the pouch and a further 1–2 outside the pouch before weaned.
DIET	A variety of leaves, particularly eucalypt leaves, fruits, buds, bark, some pasture plants
STATUS	Abundant

BRUSH-FOOTED TRAPDOOR SPIDER
Idiommata scintillans

Driving home late at night during a thunderstorm, I saw this amazing spider standing out against the white rubble driveway. It was, I presume, searching for a mate. Many male trapdoor spiders used to fall in our swimming pool during their mating seasons. This species has the ability to trap air in the fine hairs on its legs and body, which makes it fairly drown-proof compared to some spiders. Many species of trapdoor spider can revive after spending most of the night under water. A word of warning to anyone who fishes a male funnel web spider from a pool with a net. Don't leave it lying on the edge of the pool as an hour later it could bite someone if trodden upon.

COMMON NAME	Brush-footed Trapdoor Spider
SCIENTIFIC NAME	*Idiommata scintillans*
DISTRIBUTION	Throughout Australia
SIZE	3.5 cm
DIET	Frogs, lizards, geckos, large insects, spiders

AUSTRALIAN SHELDUCK
Tadorna tadornoides

Most duck species will use hollow trees as nest sites, if they are available. These shelducks won a battle with sulphur-crested cockatoos over ownership of this nest, and were able to hatch young, in spite of the competition.

The shelduck, or mountain duck as we call them, arrive each year after the opening rains. With lots of 'honking' they fight with each other over feeding areas and nest sites. Generally, they breed early in the winter.

There are many places in southern, winter-rainfall Australia where ducks breed in a short but reliable wet winter season. The water dries up in the spring, sometimes leaving many ducks 'out of water'. The survivors may move to salt swamps, the Coorong, the Murray River, or the interior.

COMMON NAME	Australian Shelduck
SCIENTIFIC NAME	*Tadorna tadornoides*
DISTRIBUTION	South-eastern NSW, VIC, south-eastern SA, western WA
SIZE	Male 59–72 cm, female 56–68 cm
BREEDING	5–15 eggs, nests in the hollow limb of a tall tree, on ground, cave floor, rabbit hole
DIET	Green freshwater plants or land plants, cereals, insects, snails
STATUS	Locally abundant

SULPHUR-CRESTED COCKATOO
Cacatua galerita

Looking out the same door, a cockatoo has just checked the contents after chasing the duck out. Fights often brewed with shelducks over nest hollows. These were grudgingly lost for the time being, as the nests would often be free again by the time the cockatoos needed them. Pairs of cockatoos regularly visit their favourite hollows throughout the year and spend long periods cuddling and preening each other. Often a group will join them to screech at interlopers. When our farm consisted mostly of scrub, there was only one pair breeding on the property. Thirty years later we had a flock of twenty-five nest sites. Cockatoos seem to have benefited from the clearing of scrub to pasture.

COMMON NAME	Sulphur-crested Cockatoo
SCIENTIFIC NAME	*Cacatua galerita*
DISTRIBUTION	Northern NT, eastern QLD, eastern NSW, VIC, TAS, south-eastern SA, Perth.
SIZE	45–52 cm
BREEDING	2 eggs, nests in hollow limb or tree hole high up in gum tree near water
DIET	Seeds of grasses and herbaceous plants, grains, bulbous roots, berries, nuts leaf buds, insects, insect larvae
STATUS	Locally abundant to common

WEDGE-TAILED EAGLE

Aquila audax

A wedge-tail stands its ground as a group of inquisitive emus approaches. A few seconds later the eagle flew at one of the emus, tearing feathers from its rump and driving the emus away from the carrion below its perch. The film in the camera was already beyond the last frame. There is nothing perfect about this photograph, but a five-foot-wide print of it resides above my desk and is a reminder of the view from the 'hide' on the hill, at home. The golden grass of early summer is at a dangerous peak for bushfire. As the summer progresses the grass will lose its colour and if not grazed will be wasted and end up as grey, washed-out straw the following winter. This land has seen the struggle for survival by animals and man for centuries.

COMMON NAME	Wedge-tailed Eagle
SCIENTIFIC NAME	*Aquila audax*
DISTRIBUTION	Throughout Australia
SIZE	Female 89–104 cm, male 87–91 cm
BREEDING	1–3 eggs, nest is a large platform of sticks in a fork of a tree, re-used year after year
DIET	Rabbits, wallabies, small kangaroos, birds, reptiles
STATUS	Common

85

OWLET-NIGHTJAR
Aegotheles cristatus

The owlet-nightjar is the nocturnal equivalent of a willie wagtail. Both display amazing agility in the air and quick movements. Their calls are often made during the day from a hollow hideaway and are commonly heard at night in many parts of the country. While I watched one night, a chocolate wattled bat entered the hollow and flew off again twenty minutes later. Another time this nightjar's nest was invaded by bees. It was immediately fumigated, however, and the nightjars were back again the following season.

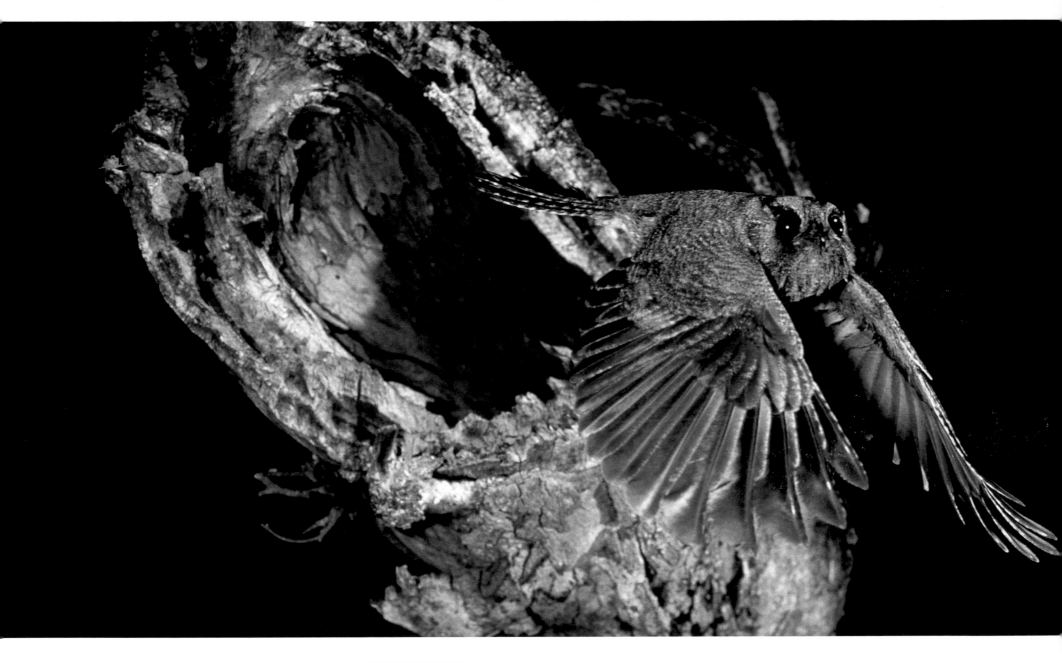

COMMON NAME	Australian Owlet-nightjar
SCIENTIFIC NAME	*Aegotheles cristatus*
DISTRIBUTION	Throughout Australia
SIZE	21–24 cm
BREEDING	2–5 eggs, nests in a tree hollow or cliff-crevice
DIET	Insects, spiders
STATUS	Common

TAWNY FROGMOUTH
Podargus strigoides

Frogmouths nested in our garden every year and the first sign of their arrival was their incessant oom-oom-oom calls, which sounded like someone in the distance starting an engine with an almost flat battery. Dung beetles, centipedes and frogs seemed to be their favourite prey. Often they would come in with a mouthful of sedge and a wet belly, indicating that they had dived into shallow water to grab a frog. Somewhere among that weed, part of the frog would be visible.

COMMON NAME	Tawny Frogmouth
SCIENTIFIC NAME	*Podargus strigoides*
DISTRIBUTION	Throughout Australia
SIZE	34–46 cm
BREEDING	1–3 eggs, in flimsy nest of twigs in fork of tree, re-uses nest yearly
DIET	Insects, spiders, centipedes, millipedes, beetles, frogs
STATUS	Common

RED FOX
Vulpes vulpes

I watched as a fox jumped a fence, crossed a road and jumped another fence before it trotted across a flooded paddock to the base of a tree on dry ground. It hesitated and then dug a small hole in the sand and appeared to bury something. When it moved off I went over and found a buff-banded rail's egg without so much as a tooth mark on it. As an experiment, I once fed some young fox cubs on our property. By using small bits of rabbit, they were enticed out of the hole under a log. I expected them to eat on the spot. Each of them took the food, trotted hundreds of metres away and buried their booty before returning to collect more and repeating the procedure. None of the food was eaten immediately.

COMMON NAME	Fox
SCIENTIFIC NAME	*Vulpes vulpes*
DISTRIBUTION	Throughout Australia except northern-most WA, NT, QLD
SIZE	Head and body length for males 61–74 cm, females 57–67 cm; tail length for males 36–45 cm, females 38–43 cm
BREEDING	4 young are born in a den
DIET	Fruits, insects, small mammals, carrion
STATUS	Abundant

FERAL CAT
Felis catus

Sadly for our native wildlife, feral cats are very successful hunters. They adapted to the bush and have proved difficult to eradicate. They catch and eat native reptiles, birds and mammals. This huge cat was feeding on a kangaroo carcass. It's amazing how much they can swallow without chewing. This must allow them to eat rapidly when competition is present. Studies of their diet prove that they consume enormous numbers of our native animals. At night, in much of the semi-desert and Mallee regions, cats are often seen in trees hunting roosting birds.

COMMON NAME	Feral Cat
SCIENTIFIC NAME	*Felis catus*
DISTRIBUTION	Throughout Australia
SIZE	Head and body length for males 45–62 cm, females 38–56 cm; tail length for males 24–34 cm, 23–32 cm
BREEDING	2 litters of 2–7 young a year
DIET	Invertebrates, fish, amphibians, reptiles, birds, small to medium-sized mammals
STATUS	Abundant

THE FIRE DRAGON

A bush fire takes hold after a lightning strike in a National Park.

Proper firebreaks are essential in parks and scrub areas. In large areas there are usually natural barriers of swamp, sand or varying vegetation that assist in the control of a wild fire. Firefighters with good local knowledge can utilise these areas, both in planning fire prevention strategies and while fighting an existing fire. Flattening scrub on its perimeter with a heavy roller provides effective reduction zones but allows a foothold for introduced weeds, as do graded tracks. Debate will continue on the best way to use fire (or whether to use it at all) as a tool towards sustainability in some areas. A study of each individual area needs to be undertaken. Australia's soils and its bush can vary so much, even within a few hundred metres.

Different plants and animals are affected by fire in their own unique way, with some benefiting and others not. The temperature of the fire decides the fate of many species. Black cockatoos, parrots and finches feed on seeds released from their pods in the fire's aftermath as some plants re-seed. Silky mice emerge from their deep sandy burrows and thrive on the fresh growth. Kangaroos return to feed on the spear grass and fresh shoots of edible plants among the ashes. Life goes on for most of the survivors and within a few short years the scrub, with short, succulent, oil-filled branches of broom bush, heath and mallee is even more inflammable than it was after twenty-five years without a fire.

Working on the dingo fence in north-eastern South Australia in the early 1960s, I witnessed thousands of rabbits moving along the northern side. They were so thick at night that it was like a nightmare. They devastated the land. When you see the damage rabbits can do to the grasses, shrubs and trees you can imagine the tonnes of insects robbed of food and shelter. These provide food for animals along the food chain, from trapdoor spiders and reptiles to birds and small mammals, including bats. Some species of plant and animal recover between the plagues but others don't.

The effect of the rabbit calicivirus since its accidental, early release in late 1995, has been dramatic. The environment will benefit even if the virus is only partially successful, particularly in the semi-desert regions. The stench of dead rabbits throughout the Flinders Ranges and far north-east region was almost unbearable, as the virus swept through. It greatly reduced rabbit numbers, even in areas experiencing above average seasonal conditions. In areas receiving below average rainfall the re-growth of plants, post-calicivirus, was substantial. A pleasing result was the sight of leaves growing on bushes well below the grazing line of the

work. There has been a trend towards responsibility for control of feral animals being worked out on the user-pays system. It is important that we realise that everyone will benefit from effective control.

The release of the calicivirus has assisted in the rehabilitation of the environment in the Yellow-footed Rock-wallabies domain by removing direct competition for food. In South Australia the Environment and Natural Resources Department has co-operated with land holders and worked to eradicate foxes both on private land and in national parks and this effort has had significant results. Rock-wallaby numbers are showing a measurable increase in these areas.

The dingo fence is a very interesting barrier. It is fascinating to see how kangaroo numbers drop off as one goes to the north of South Australia. The dingoes must have a big impact on their numbers. Only once have I seen a dingo in full flight chasing a large buck red kangaroo. It had no trouble keeping pace with it. The chase came within ten metres of the vehicle I was driving and they shot past and disappeared with the dingo right up on the kangaroo's shoulder. It may be a lot easier for dingoes to

THE INLAND

rabbit. Surveyed rabbit numbers in some places dropped to one per cent of numbers counted only months before. Some of this may have been due to normal reductions due to drought but plants thrived on deserted rabbit warrens that were completely bare a year before. The virus has not been as successful in some higher rainfall areas and further research still needs to be done.

In pastoral areas, more should be done to reduce the rabbit population. Ripping of warrens and other controls are not being undertaken as many graziers are waiting to see what happens, or cannot afford to undertake the

catch large bucks than the faster and more agile females and young animals. The big bucks tire quite quickly and turn to defend themselves.

On the southern side of the fence kangaroos are present in big numbers when the seasonal conditions are good. Kangaroos and many bird species have benefited enormously from the proliferation of water supplies installed throughout the pastoral districts. Before sheep and cattle stations were established, the red kangaroos' range would have been limited by water supply and would have contracted back to river waterholes in dry times.

RED-TAILED BLACK-COCKATOO
Calyptorhynchus banksii

This young female always wants to share her brother's meals. He is more adventurous and is always the first to start feeding. Stringy-bark seed and casuarina are the main sources of food for the black-cockatoo population along the southern border of South Australia and Victoria. The species is threatened in this area. It prefers old dead trees for its nest hollows and these are under attack from a number of quarters. Many have been cut up either because they are considered an eyesore, or for firewood. Others are now infested with termites and could collapse. European honey bees have invaded others.

Generally, this cockatoo breeds at a time of year that doesn't clash with other species, so ducks, owls, kestrels and other cockatoos can use the hollows as well. As in many other regions, the rich soils that were once covered in casuarina trees have been the first to go under the plough.

COMMON NAME	Red-tailed Black-Cockatoo
SCIENTIFIC NAME	*Calyptorhynchus banksii*
DISTRIBUTION	North-eastern WA, NT, QLD, north-eastern NSW, western WA
SIZE	63 cm
BREEDING	1 egg, nests in a tree hollow
DIET	Seeds
STATUS	Common to uncommon, depending upon locality

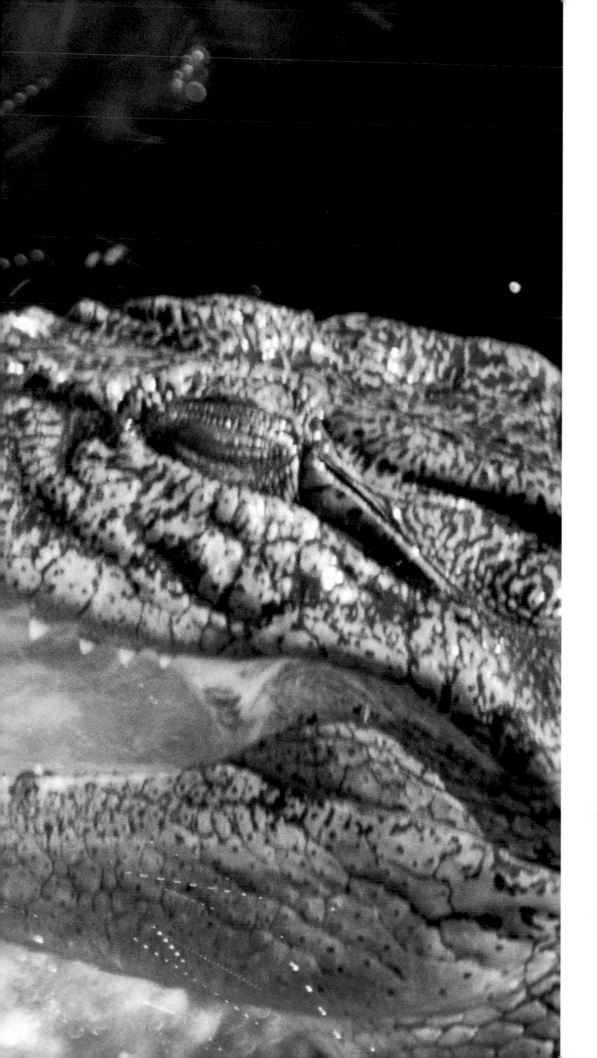

COMMON NAME	Estuarine or Saltwater Crocodile
SCIENTIFIC NAME	*Crocodylus porosus*
DISTRIBUTION	North-eastern WA, northern NT, northern and eastern QLD
SIZE	5 m
BREEDING	About 60 eggs are laid high on river banks in the wet season
DIET	Crustaceans, fish, birds, reptiles, mammals
STATUS	Common

PACIFIC BLACK DUCK
Anas superciliosa

A black duck chick was separated from the family for a short time as it hunted for food among weeds at the water's edge. It was not long before it realised the family had left it behind and its contact call became plaintive. Both parents and siblings rushed to its rescue.

Black ducks nest in many gardens in Australian cities. It is often a surprise to see a family suddenly appear on a suburban lawn, among flower beds, well away from the nearest water.

COMMON NAME	Pacific Black Duck
SCIENTIFIC NAME	*Anas superciliosa*
DISTRIBUTION	Throughout Australia except inland deserts
SIZE	47–60 cm
BREEDING	8–10 eggs, nests in scrapes in the ground, in grass, reeds or in tree holes
DIET	Seeds from aquatic plants, aquatic insects, crustaceans
STATUS	Abundant

AUSTRALIAN HOBBY
Falco longipennis

The 'little' falcon, as it's known, is like a smaller version of the powerful peregrine. A recently fledged young bird waits for its father to go hunting. One Christmas, I noticed that three young were following their male parent on his hunting forays. They confiscated all the prey he had caught before he could return to feed his mate, who had just hatched another clutch of eggs. The second lot of young died from starvation. It was the only time I've ever seen double clutching in this species. Prey, such as young starling and red-rumped parrot, were in very good supply that year.

COMMON NAME	Australian Hobby
SCIENTIFIC NAME	*Falco longipennis*
DISTRIBUTION	Throughout Australia
SIZE	Female 34–35.5 cm, male 30–32 cm
BREEDING	2–3 eggs, nests on cliff edges, in tree hollows or in the disused stick nests of other birds
DIET	Birds, bats, flying insects
STATUS	Common

YELLOW-FOOTED ROCK-WALLABY
Petrogale xanthopus

The spectacular rock faces in the Flinders and Gammon ranges of South Australia have not proved altogether safe for the rock-wallabies' continued survival. The species has been threatened with extinction by a combination of introduced animals. I have watched the wallabies travel some distance from their rocky refuge on foraging expeditions down to flatter areas where grass and herbage grow. This is often where sheep and goats graze and rabbits dig their burrows. The Environment Department is determined to eradicate goats, foxes and rabbits from these areas. The calicivirus has reduced rabbit numbers, but goats are now in demand for export so eradicating them is not a priority for many graziers.

Young wallabies, unlike kangaroos, leave their young at quite a vulnerable age among the rocks when they go off to forage. These young are easy prey to foxes. A fox-baiting program in some of the major wallaby breeding sites has been very effective. The rock-wallabies have responded in recent years by showing a measurable increase in numbers.

COMMON NAME	Yellow-footed Rock-wallaby
SCIENTIFIC NAME	*Petrogale xanthopus*
DISTRIBUTION	Flinders Ranges, SA, Barrier Range, NSW, Grey Range, QLD
SIZE	Head and body length 48–65 cm, tail length 57–70 cm
BREEDING	1 young
DIET	Grasses, plants
STATUS	Common in Flinders Ranges, rare elsewhere

99

CRESTED PIGEON
Ocyphaps lophotes

Like a 'dove of peace' a crested pigeon carries a twig towards its rickety nest high in a pink gum. This species has benefited from the clearing of land for pasture and farming in coastal districts of Australia. It is found through most of the inland, except for dry deserts.

COMMON NAME	Crested Pigeon
SCIENTIFIC NAME	*Ocyphaps lophotes*
DISTRIBUTION	QLD except for most of Cape York Peninsula, NSW, north-western VIC, SA, western and north-eastern WA
SIZE	31–35 cm
BREEDING	2 eggs, nest a frail stick platform in lower fork of bush
DIET	Seeds
STATUS	Common

SPINIFEX PIGEON

Geophaps plumifera

These plump little pigeons were photographed near Mount Meharry in Western Australia. They are often seen on the edge of roads and near water in central and northern Australia. They blend in with the red earth of the spinifex country but the detail in their plumage is exquisite when viewed at close range.

COMMON NAME	Spinifex Pigeon
SCIENTIFIC NAME	*Geophaps plumifera*
DISTRIBUTION	Western QLD, north-western SA, inland NT, north-western WA, north-eastern WA
SIZE	20–22 cm
BREEDING	2 eggs in scrape on ground
DIET	Seeds of grasses, herbs and legumes
STATUS	Common

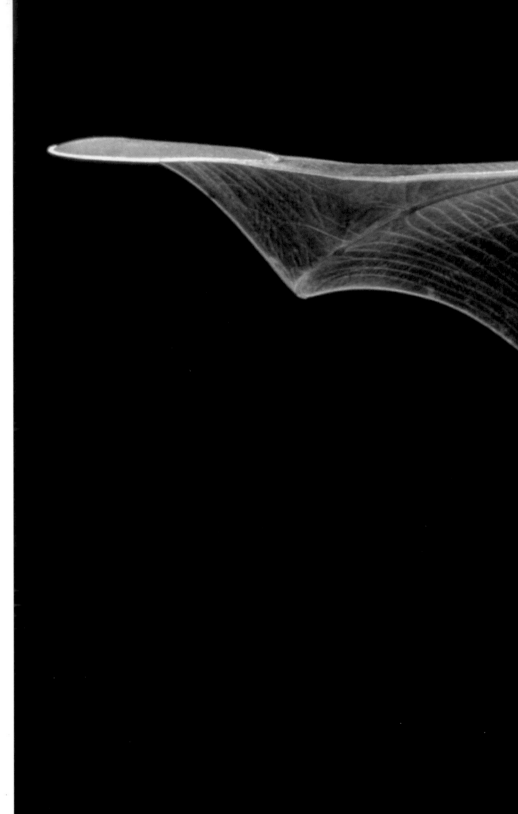

GHOST BAT
Macroderma gigas

A fantastic creature, the ghost bat has the ultimate in prey detection devices. Its massive ears allow it to pick up the sounds of prey as they move about, as well as its own echo-location sounds of varying frequencies. This species is the only carnivorous bat in Australia and catches frogs, small birds, mammals and even other smaller bats. The soft leathery skin of the wings is almost transparent. The species is threatened with extinction in much of its range and its distribution is receding due to limestone mining in some areas and intrusion by agriculture.

COMMON NAME	Ghost Bat
SCIENTIFIC NAME	*Macroderma gigas*
DISTRIBUTION	North-western WA, north-eastern WA, central-eastern WA, NT, northern QLD
SIZE	10–13 cm
BREEDING	1 young. Mothers form separate nursery colonies away from the males.
DIET	Large insects, frogs, lizards, birds, small mammals, other bats
STATUS	Sparse to rare

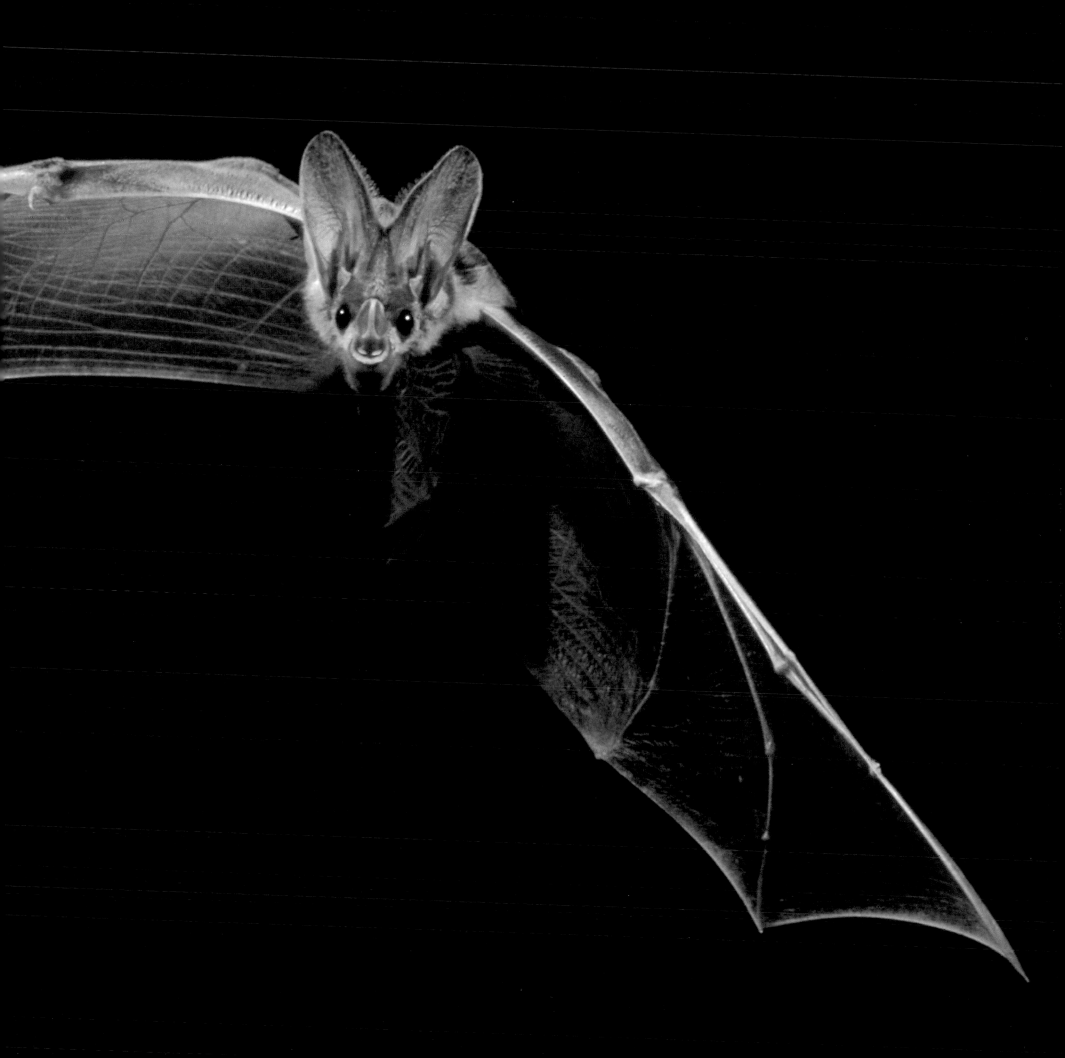

WEDGE-TAILED EAGLE
Aquila audax

I often smile as I watch the antics of young eagles in the nest. Young wedge-tails begin aerobic classes at a very young age. They exercise their wings long before feathers appear through the down. I have noticed that a high proportion of nests, in areas of reliable rainfall, fledge more than one young. The same appears to happen, even in desert areas, when conditions have been good for more than one year in a row.

Please resist the temptation to make pets of these youngsters. When they grow up they are very difficult to return to the wild successfully.

COMMON NAME	Wedge-tailed Eagle
SCIENTIFIC NAME	*Aquila audax*
DISTRIBUTION	Throughout Australia
SIZE	Female 89–104, cmMale 87–91 cm
BREEDING	1–3 eggs, nest is a large platform of sticks in a fork of a tree, re-used year after year
DIET	Rabbits, wallabies, small kangaroos, birds, reptiles
STATUS	Common

AUSTRALIAN KESTREL
Falco cenchroides

A male kestrel banks steeply in flight after sighting a mouse in the grass below. The male, with its grey head and tail, looks crisp in the early morning light against the sky. The kestrel's characteristic hovering flight can be seen along highways and roads across most of Australia.

COMMON NAME	Australian Kestrel
SCIENTIFIC NAME	*Falco cenchroides*
DISTRIBUTION	Throughout Australia except south-western TAS
SIZE	Female 33–35 cm, male 30–33 cm
BREEDING	3–7 eggs. No nest is built — it uses tree hollows, cave entrances, ledges on city buildings or cliffs, nests of raven and other birds
DIET	Insects, rodents, baby rabbits, ground birds, reptiles
STATUS	Common

MAJOR MITCHELL'S COCKATOO
Cacatua leadbeateri

The sight of hundreds of these birds filling every drinking space along the length of a 20-metre long water-trough in the Mallee is magnificent. One has to travel to Australia's comparatively dry inland areas to see this bird. At close range it is beautiful, with its pink and white body and wings, scarlet and orange crest. I once watched a group of Major Mitchells harassing a large goanna in a tree for nearly an hour near a water hole on the Darling River. The reptile ignored their complaints and just held its ground on a branch in the sun.

COMMON NAME	Major Mitchell's Cockatoo
SCIENTIFIC NAME	*Cacatua leadbeateri*
DISTRIBUTION	Central NT, southern central QLD, inland NSW, inland VIC, western SA, inland WA
SIZE	35 cm
BREEDING	2–4 eggs, nests in hollow limb or hole in tree
DIET	Seeds, nuts, fruits, roots
STATUS	Moderately common

FAT-TAILED DUNNART
Sminthopsis crassicaudata

These little creatures could become more prevalent in the dry interior if rabbit numbers were controlled. They have survived in spite of the rabbits, partly because they are carnivorous, but even their food will not survive if there is no vegetation. Many birds of prey catch this animal in spite of its mostly nocturnal habits.

COMMON NAME	Fat-tailed Dunnart
SCIENTIFIC NAME	*Sminthopsis crassicaudata*
DISTRIBUTION	Southern NT, south-eastern QLD, western NSW, western VIC, SA, southern WA
SIZE	Head and body length 6–9 cm, tail length 4–7 cm
BREEDING	8–10 young are born, but usually only 5 survive to weaning at 10 weeks of age
DIET	Insects and other invertebrates
STATUS	Common

SILVEREYE
Zosterops lateralis

The silvereye is a pest in vineyards, home gardens and orchards, raiding the fruit that has replaced the native-fruiting plants.

While watching Australian hobbies hunting, I noticed on a number of days that the male was flying steeply into the sky to catch very small prey. The birds were so high that I couldn't even see them with very powerful binoculars, but after a number of rapid stoops the hobby returned each time clutching a dead silvereye. I wondered why the tiny birds were so high up? The high altitude may allow them to use jet stream like wind currents on their long migrations. These tiny birds migrate across Bass Strait and travel as far as Queensland and out to offshore islands.

COMMON NAME	Silvereye
SCIENTIFIC NAME	*Zosterops lateralis*
DISTRIBUTION	QLD coast, eastern NSW, VIC, TAS, southern SA, southern and south-western WA
SIZE	12 cm
BREEDING	2–4 eggs, small nest on outer branch of tree
DIET	Insects, nectar, seeds, fruit
STATUS	Common

CAPE BARREN GOOSE
Cereopsis novaehollandiae

Cape Barren goose bathing. Kangaroo Island would not have been very hospitable to these birds before settlement began, with the food supply limited to a few tidal and river flats and lakes. The birds were introduced and thrived, along with pasture development in the Flinders Chase area. With the advent of irrigated pastures on the Lower Murray region, flocks of many hundreds of birds visit the area to feed. The geese are a magnificent sight as they wing their way from their feeding grounds to water, flying in skeins across the skyline.

COMMON NAME	Cape Barren Goose
SCIENTIFIC NAME	*Cereopsis novaehollandiae*
DISTRIBUTION	Islands off southern coast of Australia
SIZE	75–100 cm
BREEDING	1–7 eggs, breeds May–August
DIET	Grasses, sedges, clovers, herbs, seeds
STATUS	Moderately common

YELLOW-FOOTED ROCK-WALLABY
Petrogale xanthopus

A rock-wallaby feeding a joey that will soon be independent. Now, it is still drinking milk from its mother's pouch and is left for quite long periods on its own among the rocks to fend for itself between her visits. The mother may go some distance away to feed, depending on local conditions.

COMMON NAME	Yellow-footed Rock-wallaby
SCIENTIFIC NAME	*Petrogale xanthopus*
DISTRIBUTION	Flinders Ranges, SA, Barrier Range, NSW, Grey Range, QLD
SIZE	Head and body length 48–65 cm, tail length 57–70 cm
BREEDING	1 young
DIET	Grasses, plants
STATUS	Common in Flinders Ranges, rare elsewhere

SOUTHERN BOOBOOK
Ninox novaeseelandiae

Two young boobook owls sit at the entrance to their nest hollow waiting to be fed. Their heads constantly turn as they watch flying insects whizzing past on a warm summer evening. Their parents had been catching mostly dung beetles for their young but suddenly one arrived with a small brown bat. The following week the hollow was taken over by bees and the young, who could just fly, moved to roost in a clump of wattle trees.

COMMON NAME	Southern Boobook
SCIENTIFIC NAME	*Ninox novaeseelandiae*
DISTRIBUTION	Throughout Australia
SIZE	30 cm
BREEDING	2–4 eggs, nest is a tree hollow lined with wood chips, leaves and twigs, breeds August–January
DIET	Small birds, small mammals, beetles, moths
STATUS	Common

MALLEEFOWL
Leipoa ocellata

It was the stroke of midday on New Year's Day. The temperature was 42°C. After sitting in a hide overlooking the nest from an hour before sunrise every day for some time, at last I was rewarded by the hen laying an egg.

It was fascinating watching the birds' behaviour over many weeks. After prolonged periods of digging in the hot sun the male would often retire to the shelter of the nearby scrub. He would quickly dig a shallow hole in the soil beneath the shade and stand on the cooler, fresh layer of earth. Sometimes he would lie on his side and stretch his legs. Both birds were aware of eagles and hawks in the sky. Excavation of the mound would stop and the birds would look up with their heads cocked on one side. Invariably, following that gaze with field glasses would reveal an eagle high above. If the eagle became visible to the naked eye one, or both birds, would quickly retire to the shade. Once a brown falcon flew low overhead and surprised both birds on the mound. A loud alarm call accompanied their panic-stricken dash for cover, although they probably had little to fear from a falcon of that size.

Five nests were active in the scrub reserves on the property and provided many fascinating moments. Sleepy lizards would regularly visit the mounds to eat the dung of the malleefowl, which must have contained some nutrient or moisture. Small white-browed scrubwrens, emuwrens, and blue wrens flitted around in the nearby scrub. It is amazing how many varieties of birds can be seen when you're sitting quietly, compared with either walking or driving in the bush. These are some of the reasons for leaving areas of scrub unscathed by bulldozer, plough or livestock.

COMMON NAME	Malleefowl
SCIENTIFIC NAME	*Leipoa ocellata*
DISTRIBUTION	Central NSW, western VIC, eastern SA, central SA, inland south-western WA
SIZE	60 cm
BREEDING	5–33 eggs are incubated in a large mound of vegetation covered with sand.
DIET	Acacia seeds, buds of herbs, insects
STATUS	Uncommon

GUM-TREE MOTH LARVAE
Uraba lugens

Feeding as a mob on gum trees, these caterpillars skeletonise the leaves. Many insects make up part of the total ecosystem on any one species of tree and recycle some of the nutrients through their droppings to the tree roots below. They, in turn, provide food for other insects, reptiles and birds. The caterpillars create intriguing patterns in the leaves as they graze and move. Their bodies are well camouflaged by design and colour. With clever use of their feet some birds eat the inside portion of these grubs and avoid the irritating hairs.

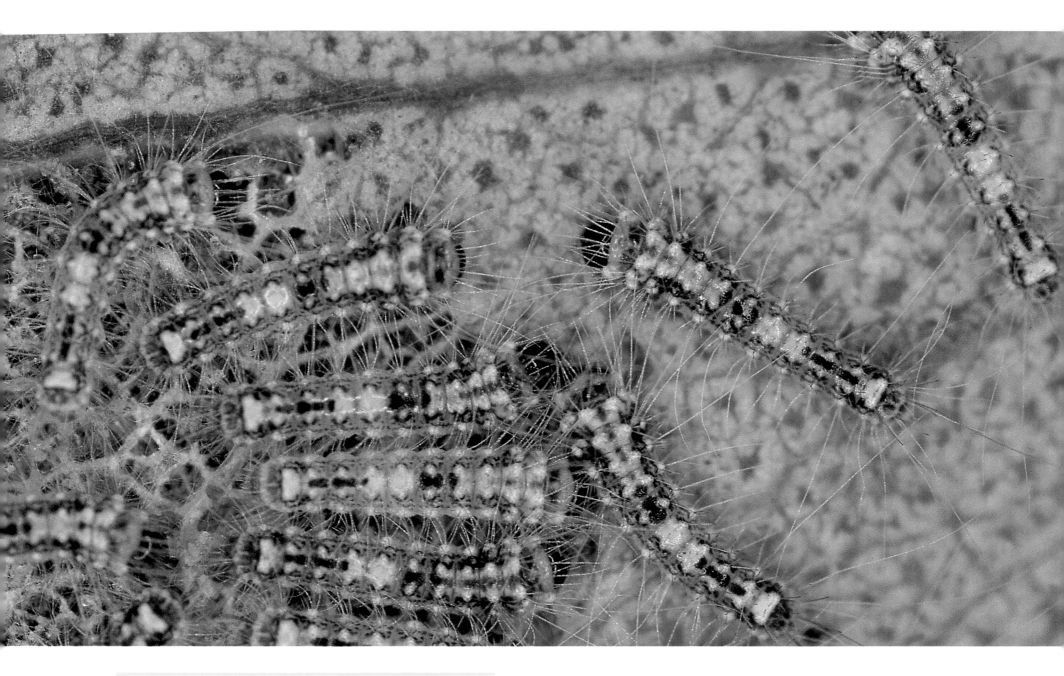

COMMON NAME	Gum-tree Moth
SCIENTIFIC NAME	*Uraba lugens*
DISTRIBUTION	Coastal north-eastern QLD, south-eastern QLD, NSW, VIC, TAS, southern SA, south-western WA
SIZE	2.5 cm
DIET	Caterpillars feed on eucalypt leaves
STATUS	Common

FERAL (EUROPEAN) HONEY BEES
Apis mellifera

This tree hollow is only small but valuable nonetheless. A colony of Gould's wattled bats roosted and reared their young only the year before.

Over 160 feral hives were found on my property and at least thirty per year were fumigated repeatedly over twenty-five years. It usually took two years before the hollows were re-used by their rightful tenants. Hollows of all sizes, used by ducks, owls, kestrels and cockatoos down to small parrots, pardalotes and bats were taken over. Often the bees would move into a small hollow that had no hope of supporting them and honeycomb would grow from its entrance only to melt in the sun or to be blown away, leaving an abandoned hive blocking the entrance.

COMMON NAME	Honey Bee
SCIENTIFIC NAME	*Apis mellifera*
DISTRIBUTION	Throughout Australia
SIZE	1-2 cm
BREEDING	The queen bee lays thousands of eggs, which grow into worker bees, who tend the queen, eggs and young.
DIET	Nectar
STATUS	Common

117

DWARF COPPERHEAD SNAKE
Austrelaps labialis

'Tasting the air' with its tongue, this small snake hesitated at the sight of the camera before moving quickly across a flooded creek to a hole in some rocks on the far bank. It had been hunting for small lizards among the rocks beneath a blue gum. As I climbed down from my hide, which was set up next to a koala, I nearly trod on it. Dwarf copperheads are seen quite often on Kangaroo Island. I remember stopping the traffic for five minutes on the way in to Cape Du Couedic to allow one to safely cross the road. Like the black tiger snakes that inhabit the island they are often seen near the coast.

COMMON NAME	Dwarf Copperhead
SCIENTIFIC NAME	*Austrelaps labialis*
DISTRIBUTION	Kangaroo Island and southern Mt Lofty Ranges, SA
SIZE	84 cm
BREEDING	2–10 young
DIET	Frogs, small mammals, reptiles
STATUS	Common

119

The rescue and rehabilitation of injured wild animals and birds seems to have become a real growth industry in Australia in recent years. Many birds are rescued unnecessarily. Some rescuers are aggressive in their attempts to make pets of almost any wild animal or bird and end up with collections of lame wildlife. Many young birds leave their nest long before they can fly, with short tails and stumpy wings. Young magpies, thrushes, wrens and many others set off into the world on foot. Many fall prey to snakes, goannas and hawks. Young ducks sometimes have to walk for a long time before they reach water or learn to fly. Birds are injured or killed when they hit power lines, fences, windows and motor vehicles.

There are now hundreds of injured birds in captivity but I believe that most of these birds should have been put down. Most rehabilitated birds and animals fail to survive back in the wild. There may be some value in allowing people the privilege of observing these wild creatures' habits, but collecting becomes an obsession with some people. It is a poor substitute for watching wildlife in the wild. People who keep solitary birds in captivity are privy to such behaviour as a pet cockatoo preening its human captor as it would its mate. They may treat a human as one of 'their own', but I don't think birds and animals 'adopt' human feelings.

Investing huge resources in rescuing rare and endangered wildlife will be a waste of time unless the habitat in which they live is preserved.

Wildlife photography

Some wild animals and birds are very sensitive to disturbance, but if care is taken, combined with knowledge of the species' behaviour, it is possible to study their habits and photograph them without doing irreparable harm. Lack of time, lack of determination, or lack of patience are the biggest enemies of ethics. Even with time, determination and patience in abundance, a photographer still needs a lot of luck for the perfect shot.

Care should be taken at all times. Nest trees should not be cut into just to get a photograph from an unusual angle. Nests should not have branches broken which may impede the view from the camera.

It's not smart to be the only one with a photograph or movie of some rare species if it has terminated their breeding efforts for the year. You should never annoy or harass wildlife unnecessarily. The best approach is to learn from quiet observation and let the photograph be of secondary consideration. There are times when you could surprise a bird with a radio-controlled camera and get a photograph, but it may be impossible to get another photo of the same bird. I would never leave a camera, or any equipment, attatched to or close to a wedge-tailed eagle nest. Imagining yourself in the position of the subject is not a bad idea. And sometimes it is simply better to settle for photographing some species' behaviour in captivity than to risk disturbing wild birds and animals.

TO THE RESCUE

ZEBRA FINCH
Taeniopygia guttata

A male finch sits in a boxthorn bush close to the nest he has helped to build in its protective branches. Many small birds use this feral bush for their nests as it has colonised watercourses and the edges of man-made dams in outback Australia. Zebra finches often build their nests at the base of raven and eagle nests where they are protected. When drinking from waterholes in hot weather they gather in large numbers and shelter in shrubs before venturing to the water's edge. The black falcon, although capable of taking quite large prey, at times shows a preference for these little birds.

COMMON NAME	Zebra Finch
SCIENTIFIC NAME	*Taeniopygia guttata*
DISTRIBUTION	Throughout Australia except Cape York Peninsula
SIZE	10 cm
BREEDING	4–5 eggs, nest of grass-stems in twigs of bushes and low trees, sometimes in hollows or on the ground
DIET	Fallen grass seeds
STATUS	Common to uncommon

WHITE-FACED HERON
Egretta novaehollandiae

Nesting within five metres of a peregrine falcon would seem to be suicide for a bird known to be preyed on by its neighbour, but these two species lived in comparative harmony. A yellow-billed spoonbill also nested nearby. The herons often visited the carcasses of dead sheep to catch brown blowflies, which they deftly snapped up with pinpoint accuracy. Small fish, frogs, tadpoles and insects form most of their diet. Herons frequently nest close to farmhouses and are very noisy nest builders.

Note the unhatched egg still in the nest.

COMMON NAME	White-faced Heron
SCIENTIFIC NAME	*Egretta novaehollandiae*
DISTRIBUTION	Throughout Australia except central WA
SIZE	60–70 cm
BREEDING	Up to 7 eggs, nest is a loose platform of sticks in a tree
DIET	Crustaceans, squid, fish, insects, amphibians, spiders, snails, worms
STATUS	Common

BLACK-FACED WOODSWALLOW
Artamus cinereus

Nest building in the split trunk of a dead coolabah tree on the Strzelecki Creek. When in flight, woodswallows have a certain grace that is unique among Australian birds. With their plump, but streamlined bodies, guided by tails that continually correct for wind currents, their wings pump furiously one second and hold a smooth glide the next as they hawk for insects high in the sky.

COMMON NAME	Black-faced Woodswallow
SCIENTIFIC NAME	*Artamus cinereus*
DISTRIBUTION	Throughout Australia, except Cape York Peninsula, coast of NSW and VIC, south-eastern coast SA, south-western coast WA
SIZE	18 cm
BREEDING	3–4 eggs, nest a sparse cluster of twigs in a bush, stump or fence-post
DIET	Insects
STATUS	Common

YELLOW-TAILED BLACK-COCKATOO
Calyptorhynchus funereus

Pine trees provide food and shelter for a number of bird species. Large flocks of black-cockatoos fly for many kilometres between plantations. Pine trees supply a reserve of food that may fill gaps in availability of their natural food. They often carry cones some distance from the trees, which can lead to the spread of feral pines. This has happened on Kangaroo Island, in the Adelaide Hills and Hindmarsh Island in South Australia. Other parrot species eat the seeds and lorikeets feed on the pollen of these trees.

COMMON NAME	Yellow-tailed Black-Cockatoo
SCIENTIFIC NAME	*Calyptorhynchus funereus*
DISTRIBUTION	Southern QLD, eastern NSW, VIC, TAS, south-eastern SA
SIZE	56–66 cm
BREEDING	2 eggs, nests in a very high tree hollow
DIET	Seeds, insect larvae
STATUS	Common

SUGAR GUM FOREST

This is typically degraded sugar gum country, where livestock have grazed out the understorey. No re-growth of young replacement trees or shrubs and sedges will be possible as long as livestock have access. This is a favoured nesting habitat for glossy black-cockatoo as well as other bird species. The mature sugar gum contains a reasonable number of hollows but as time passes, this patch will die out. The trees were left for shelter and for their beauty but are doomed eventually without a good stock-proof fence. The area, with the invasion of grass and the aid of fire, will eventually become an open, pastured hillside. Thousands of hectares across Australia are in the same predicament. It is cheaper to protect existing areas of bush now with fencing, than to plant millions of trees later. Low wool and commodity prices make it hard for farmers to pay for the fencing now but it should be undertaken immediately. The message is simple: livestock and native vegetation are not compatible.

127

BEAUTIFUL FIRETAIL
Stagonopleura bella

A male firetail feeds on seed from fallen casuarina cones. At close range this is a stunningly beautiful little finch, which I associate with many hours spent watching the Glossy Black-cockatoo on Kangaroo Island. This species, along with many others, depends on the casuarina for at least part of its diet. The soil on which the black oak thrives has been one of the first to be cleared of native vegetation for farming wherever it grows in Australia. It has been easily eliminated with repeated burning and grazing with livestock and it's sad that more reserves were not left.

COMMON NAME	Beautiful Firetail
SCIENTIFIC NAME	*Stagonopleura bella*
DISTRIBUTION	South-eastern NSW, southern VIC, TAS, south-eastern SA
SIZE	11–12 cm
BREEDING	2–6 eggs, nest is a bulky dome in shrubbery near the ground
DIET	Grass seeds, sheoak seeds, flower petals
STATUS	Uncommon

MALLEE EMUWREN
Stipiturus mallee

This bird was photographed in the garden within two years of planting a corridor of trees and shrubs linking the house paddock scrub on our property with larger areas of bush in reserves further away. This species had not been seen in the area for twenty years. It demonstrates the value of corridors in allowing species to move around, thereby increasing the diversity within species as well as variety. The same corridor is now used regularly at dusk, and even after dark, by thousands of honeyeaters as they move from feeding areas to roosting places some kilometres apart.

COMMON NAME	Mallee Emuwren
SCIENTIFIC NAME	*Stipiturus mallee*
DISTRIBUTION	South-eastern SA/VIC border
SIZE	16.5 cm
BREEDING	2–3 eggs, nest a small ball of interwoven fibres
DIET	Insects, seeds
STATUS	Moderately common to uncommon

GREY CURRAWONG

Strepera versicolor

As I watched through a telescopic lens, the gory details of feeding time were suddenly revealed to me. The parent currawong had flown a kilometre across open country carrying food from an isolated patch of scrub. The food was a very lively centipede, which was shoved down the throat of the first baby to open its mouth. I shuddered when the centipede proceeded to crawl out of the baby's throat and fell into the nest. Undeterred, the parent quickly picked it up and popped it down the hatch again. The determined centipede repeatedly escaped only to be replaced again, and again. Eventually one of the young presented a little parcel of waste which distracted the parent just long enough for the centipede to escape.

COMMON NAME	Grey Currawong
SCIENTIFIC NAME	*Strepera versicolor*
DISTRIBUTION	South-eastern NSW, VIC, TAS, southern SA, south-western WA, central Australia
SIZE	45–53 cm
BREEDING	2–3 eggs, nest a shallow cup of twigs in a tree fork
DIET	Insects, small vertebrates, eggs, nestlings, berries
STATUS	Common

131

SHIELD SHRIMP
Triops australiensis

These tadpole-sized creatures, looking like extinct trilobites, are adapted to temporary fresh water pools and claypans. They are found across most of the inland including the Great Sandy Desert. Their eggs are drought tolerant and are blown around by wind. They seem to materialise in a matter of days after rain and rush through their life cycle in the race against time, before being desiccated by the sun and wind again. They are an inconspicuous part of desert life, hidden by the clay particles in water, until evaporation exposes them in their millions.

COMMON NAME	Shield Shrimp
SCIENTIFIC NAME	*Triops australiensis*
DISTRIBUTION	Throughout Australia except the south-west, south-east and far north
SIZE	9 cm

GREEN TREE-FROG
Litoria caerulea

Green tree-frogs often confuse the sound of human voices with their own mating calls and join in with their own chorus. The smug look on this frog's face was worth preserving. It may be the result of surviving a long trip from Queensland to Adelaide in a bunch of bananas.

COMMON NAME	Green Tree-frog
SCIENTIFIC NAME	*Litoria caerulea*
DISTRIBUTION	North-eastern WA, north-eastern NT, QLD, NSW (except extreme south), north-eastern SA
SIZE	10 cm
BREEDING	200–2000 eggs per breeding season
DIET	Insects, small birds, rodents
STATUS	Abundant

WILD GOATS IN THE
FLINDERS RANGES

At the centre of deep blue shade, the early morning light on a
mountain face is reflected in a pool of the Frome Creek, in the
north Flinders Ranges in South Australia. A small herd of feral
goats comes down to drink and then disperses among the rocks
where a colony of yellow-footed rock wallabies is trying to eke
out a living. The goats stop to browse on almost every plant the
wallabies treasure and sometimes stand on their hind limbs to
eat leaves and bark high up in shrubs. Although wild, the goats
are still being 'farmed' by landholders, who need the income
from their sale to survive on the land, while wool prices and
returns from cattle are low. After picking the big, saleable goats
from mobs, young goats are often released to grow and breed
again. This makes the total eradication of the feral goat
impossible.

LITTLE CORELLA
Cacatua sanguinea

Late evening at a waterhole along the Strzelecki Creek. Many hundreds of corellas congregate at these waterholes to spend the night near the water in the shelter of the coolabahs. Often on moonlit nights in the late spring, the corellas call and feed their young during much of the night.

COMMON NAME	Little Corella
SCIENTIFIC NAME	*Cacatua sanguinea*
DISTRIBUTION	North-western WA, north-eastern WA, NT, western QLD, western NSW, western VIC, eastern SA
SIZE	36–39 cm
BREEDING	2–3 eggs, nests in tree hollow
DIET	Seeds of grasses and legumes
STATUS	Common to locally common

BLACK-BROWED ALBATROSS

Diomedea melonophris

In the Southern Ocean, out from the Great Australian Bight, a young albatross comes in with shearwaters to feed on off-cuts of bait from fishing traps. In flight, with strong winds, this is a magnificent bird to watch as it dips in and out of the swell, its wing tips almost touching the waves.

COMMON NAME	Black-browed Albatross
SCIENTIFIC NAME	*Diomedea melanophris*
DISTRIBUTION	Off coasts of southern QLD, NSW, VIC, TAS, SA, southern WA
SIZE	85 cm, wingspan 240 cm
BREEDING	1 egg, breeds in colonies on bank, ledge or slope
DIET	Squid, crustaceans, fish
STATUS	Locally abundant to moderately common

LITTLE CORELLA
Cacatua sanguinea

With wings spread wide in a threat display and loud screeching, a corella announces the presence of a large brown snake that has just slithered out of the waters of the Cooper Creek beneath its nest. Brown snakes climb quite large trees to rob birds nests of their eggs and young.

COMMON NAME	Little Corella
SCIENTIFIC NAME	*Cacatua sanguinea*
DISTRIBUTION	Northern WA, NT, QLD except Cape York, NSW into eastern SA, northern VIC
SIZE	35–38 cm
BREEDING	3–4 eggs, nests in holes in trees, sometimes cliffs
DIET	Seeds, but also fruit, roots and insects
STATUS	Abundant within range

NEST ROBBERS

This nest hollow has been plundered by human poachers cutting in through its base to get at fledglings, eggs or reptiles and is a disheartening sight. The hollow is now useless for birds or animals. This sight is all too common in much of remote, inland Australia. The vast inland creek systems lined by eucalypts are raided by many smugglers. Many of their captive animals die from lack of air or maltreatment.

BEARDED DRAGON
Pogona vitticeps

In the southern winter rainfall area, this reptile takes on a range of colours that suit the locality. These creatures often control their own body temperature and survey their territory from a high position on a post or tree. Much territorial head-bobbing accompanies close encounters such as the one I caught on camera here.

COMMON NAME	Bearded Dragon
SCIENTIFIC NAME	*Pogona vitticeps*
DISTRIBUTION	Inland QLD, inland NSW, north-western VIC, north-eastern SA, south-eastern NT
SIZE	20 cm
BREEDING	1 or 2 clutches of 11–25 eggs
DIET	Insects
STATUS	Common

AUSTRALIAN WHITE IBIS
Threskiornis molucca

Nest building is on in earnest at a breeding colony of white ibis, at Bool Lagoon in South Australia. After a colony is established, the ibis show remarkable loyalty to the site, remaining in the area and gradually building their numbers if food is available. They have a tendency to crush plantations of trees in which they nest.

Ibis flock to ephemeral waterholes as these dry out in late spring, to feed on tadpoles and newly emerging froglets. This bird consumes millions of budworm and other pests in pastures and crops. The ibis is a symbol for conservation in South Australia.

COMMON NAME	Australian White Ibis
SCIENTIFIC NAME	*Threskiornis molucca*
DISTRIBUTION	NT, QLD, NSW, VIC, TAS, eastern SA, south-western WA, north-western WA
SIZE	65–75 cm
BREEDING	2–5 eggs, nests on a platform of sticks
DIET	crustaceans, water insects, fish, snails, frogs, crickets, earthworms, small snakes
STATUS	Locally abundant to common

143

PEARSON ISLAND ROCK-WALLABY

Petrogale lateralis pearsoni

A young wallaby licked its mother's mouth for twenty minutes as she regurgitated water or food for it. Earlier the mother had taken thirty minutes to glean drink from a small spring in rocks nearby. This behaviour has also been noted in other rock-wallabies.

COMMON NAME	Pearson Island Rock-Wallaby
SCIENTIFIC NAME	*Petrogale lateralis pearsoni*
DISTRIBUTION	Pearson Island, SA
SIZE	Head and body 45–53 cm, tail 48–60 cm
BREEDING	1 young
DIET	Grasses
STATUS	Rare

RED KANGAROO
Macropus rufus

The red kangaroo prefers wide open spaces but seeks shade when it's hot. A week after a heavy thunderstorm I witnessed an amazing migration of red kangaroos. The prevailing westerly carried the scent of damp soil and green grass, drawing the kangaroos from as far as the eye could see. They were all moving in the same direction, and continued to come from the east for two days. I noticed how much less agile this species is than its grey relations. They often tripped over the tops of salt-bush and blue bush plants, unable to protect themselves very well with their spindly forelegs as they crashed to the ground.

COMMON NAME	Red Kangaroo
SCIENTIFIC NAME	*Macropus rufus*
DISTRIBUTION	Inland Australia and on coast where rainfall less than 50 cm a year
SIZE	Head and body length for males 93–140 cm, females 74–110 cm; tail length for males 71–100 cm, females 64–90 cm
BREEDING	2–3 litters a year, of up to 6 young
DIET	Grasses, plants
STATUS	Abundant

145

WEDGE-TAILED EAGLE
Aquila audax

Imagine being an emu chick with this wedge-tail in pursuit. Wedge-tails will sometimes defend their nests in quite terrifying fashion. A number of pairs, whose progress I have regularly checked, greet me on the first visit of the year by dive bombing. They come from high in the sky, just straight at me and 'swoosh' overhead. Sometimes they beat their wings or breast against leafy branches to reinforce their effort. The hair stiffens on the back of my neck. It's difficult to hold your nerve in this situation and to keep the camera still!

COMMON NAME	Wedge-tailed Eagle
SCIENTIFIC NAME	*Aquila audax*
DISTRIBUTION	Throughout Australia
SIZE	Female 89–104 cm, male 87–91 cm
BREEDING	1–3 eggs, nest is a large platform of sticks in a fork of a tree, re-used year after year
DIET	Rabbits, wallabies, small kangaroos, birds, reptiles
STATUS	Common

CHESTNUT TEAL

Anas castanea

The male duck is brilliantly coloured, while the female's mottled, grey-brown and speckled appearance is beautiful in its own right. These birds regularly used artificial hollows erected in the bush. The male took quite an interest in rearing his offspring, even sometimes feigning a broken wing if anyone ventured too close to his family.

COMMON NAME	Chestnut Teal
SCIENTIFIC NAME	*Anas castanea*
DISTRIBUTION	South-eastern QLD, NSW, VIC, south-eastern SA, south-western WA
SIZE	38–48 cm
BREEDING	7–10 eggs, nest is a scrape in the ground, in long grass, crevices in rocks or tree holes close to the water
DIET	Grasses, seeds, snails, crustaceans, insects, worms
STATUS	Abundant to locally common

GREY TEAL
Anas gracilis

When the normal winter rainfall occurs in southern Australia, grey teal ducks start nest-hunting in early spring. Competition is fierce for suitable hollows and fights between pairs often occur, high up in trees. The female's loud slightly maniacal chuckle is often heard through the night. Huge flocks of grey teal gather on larger lakes as water recedes in late spring and early summer.

COMMON NAME	Grey Teal
SCIENTIFIC NAME	*Anas gracilis*
DISTRIBUTION	Throughout Australia
SIZE	37–48 cm
BREEDING	4–14 eggs, nests on ground, in rabbit burrows, crevices in rocks, or in tree holes
DIET	Seeds of plants in or overhanging water, insects, mussels, crustaceans
STATUS	Abundant

149

CLAY PAN IN THE GREAT SANDY DESERT

During the wet season, clay pans can be 'dry as a chip' one day and swimming holes the next. Life springs up at an amazing rate. Frogs emerge the night after rain and join campers around a campfire, coming in close to help themselves to insects, drawn to the light of the flames. Water birds appear and the desert blooms within weeks. Dingoes, freed from limited water supplies, have the opportunity to follow their noses across the desert, their tails flagging their progress above the vegetation.

151

Most animals and birds make some use of camouflage. If they are not cryptically marked or coloured themselves, they will take cover in places that hide or match their shape. The one thing that gives them all away is movement. A kestrel can see a mouse or a grasshopper move from hundreds of metres away but has little chance of seeing it if it doesn't move. Many animals' first instinct is to keep still when threatened. Even the hunters will do the same when the tables are turned and they become the hunted. A grey falcon will freeze on the spot while feeding in open country if a black falcon or buzzard flies overhead. With its head cocked, it will scan the sky with one eye and keep still until the danger is past. Reptiles such as the bearded dragon, gecko and frog can change colour. Lizards also rely on freezing on the spot as a defence. The smaller skinks and dragons move very rapidly between locations in fits and starts, looking this way and that with their eyes between forays.

Trapdoor spiders have an amazing array of openings to their burrows. Some are just open-topped holes, but others have tightly fitting lids of varying thickness. When the lid is shut the entrance is impossible to pick

DON'T MOVE OR YOU'RE DEAD

from the surrounding soil. Some burrows have a silken sock-like entrance which collapses, leaving only a slight subsidence in the sand as evidence of its presence. Some holes are decorated by twigs that form a pattern rather like a necklace, but it may confuse predators or warn of approaching prey.

Some insects also change colour, and have evolved over generations to match the plants on which they feed. The same thing applies to a photographer in a hide. No matter how good the camouflage, if a lens is moved or the material blows in the wind the subject will see it and be distracted or frightened.

GRASSHOPPER
Family Acrididae

A small green grasshopper almost matches the leaves of a wild tomato. A small furry caterpillar was also found on the bush and both were difficult to locate because they were so well camouflaged.

BLACK SWAN

Cygnus atratus

The sight of cygnets riding among the curly feathers on a parent's back is a delight. By contrast, territorial battles waged over patches of water by parent birds at nesting time can be spectacular. They rush across the water at intruders, legs sprinting along the surface and wings flapping. While watching nocturnal creatures well inland, I have often heard swans call as they move cross-country at night. The sound is a reminder of the long journeys many water birds make, flying steadily at night, without the use of thermals to help them on their way.

COMMON NAME	Black Swan
SCIENTIFIC NAME	*Cygnus atratus*
DISTRIBUTION	South-eastern QLD, NSW, VIC, TAS, southern SA, southern and western WA
SIZE	106–142 cm, wingspan 160–200 cm
BREEDING	2–3 litters a year, of up to 6 young
DIET	Aquatic plants
STATUS	Abundant to common

EASTERN ROSELLA
Platycercus eximius

A male rosella, stretching his tail and wing while sitting near his nest.

There were a number of artificial nest hollows in our garden at home, but only one pair of rosellas nested successfully each year. The competition for the hollows was intense. Mallee ringnecks and rosellas would fight both within their species and between species over suitable hollows. Often, with loud chittering noises, there would be up to six birds locked in a cloud of colour in mid-air. Beautiful feathers would float to earth after a few seconds, while the birds retired to their corners to regroup. During the late winter, we would often have to remove starlings' nests from most of the hollows. Then, just as the young parrots were almost ready to leave the nests, feral bees would start swarming, often moving in to start a new hive in hollow trees filled with baby birds or eggs.

COMMON NAME	Eastern Rosella
SCIENTIFIC NAME	*Platycerus eximius*
DISTRIBUTION	South-eastern QLD, along NSW border, eastern NSW, VIC (except far west), eastern TAS
SIZE	30 cm
BREEDING	4–6 eggs, nests in a deep hole or hollow in a tree
DIET	Seeds, fruit, blossom, nectar, insect larvae
STATUS	Common

155

BROLGA

Grus rubicundus

Brolgas silhouetted against the setting sun as they fly down the Frome Creek towards Lake Eyre in early summer. It is always a pleasant surprise to see a pair in the dry interior, where recent rain provides an opportunity for them to roam new pastures.

COMMON NAME	Brolga
SCIENTIFIC NAME	*Grus rubicundus*
DISTRIBUTION	Inland Victoria and NSW, northern NSW, QLD, north-eastern corner of SA, north and east NT, north WA
SIZE	Up to 140 cm
BREEDING	1–3 eggs, nest is a raised mound of coarse grass, sticks and leaves, above water or on dry land
DIET	Tubers of sedges, grain, snails, insects
STATUS	Common

156

FRIEND OR FOE

A revolution is taking place in the way chemical sprays are used on farms. Lessons have been learned from experience, both in Australia and overseas about over-use of chemicals, which has hastened the development of resistant strains of disease and pests. Scientific monitoring is used to determine the time to spray for best effect or whether in fact spray is needed at all. There are strict controls on types of chemicals and their uses as well as prohibition of long lasting pollutants that can damage the environment. It is still disappointing to watch bats flying at night, in the landing lights of a plane, while it sprays the very insects that the bats are eating. A continual effort is needed to find safe biological control methods.

DO TREES GROW UPSIDE DOWN IN THE DESERT?

The Strzelecki Desert sands are constantly on the move. The winds can pick up the soil and unearth a tree, then send it cart-wheeling across the desert and transplant it upside down. More sand anchors it for a time. Rabbits and cattle denude the soil, probably contributing to this process.

WEDGE-TAILED EAGLE
Aquila audax

A young eagle approaches for a graceful and controlled landing into the prevailing wind. A mature pair of black wedge-tailed eagles tolerate a number of young birds of different ages within their nesting territory. They often allow up to ten to hunt and feed quite close to their nest. A mature bird is a different story. Usually an intruder will heed the body language of the resident pair, patrol-flying above their territory, and shy away from the area altogether. The residents spend a number of hours a day, gliding along ridges displaying their ownership, with wings held outstretched. They rock from side to side on the wind. Rising quickly on thermals in the heat of the day, they are soon almost invisible to the naked human eye.

I watched recently as a pair of eagles rose quickly on a thermal in the heat of the day. They were soon so high they were almost invisible to the human eye. Their decent was spectacular as both birds fell at great speed across the sky with their wings swept back. They pitched forward into a much steeper dive, swooping down and then up to the point of stall. They hung, in turn, for a moment as if hooked by their beaks, with wings folded. They plummeted again and again, following each other in a series of wonderful stoops and stalls. They eventually landed on top of a bullock bush near their nest and with loud yelping calls the male mounted the female and they mated.

COMMON NAME	Wedge-tailed Eagle
SCIENTIFIC NAME	*Aquila audax*
DISTRIBUTION	Throughout Australia
SIZE	Female 89–104 cm, male 87–91 cm
BREEDING	1–3 eggs, nest is a large platform of sticks in a fork of a tree, re-used year after year
DIET	Rabbits, wallabies, small kangaroos, birds, reptiles
STATUS	Common

PEREGRINE FALCON
Falco peregrinus

A male peregrine adopts a slightly threatening posture as his mate lands above, on top of the hide. This photograph was taken at such close range, with a feeling of great anxiety, using a wide-angle lens. The shutter seemed to go off with unbearable noise but luckily the wind and the distraction of his mate's presence kept the bird calm. Working as close as this to a wild bird is a thrilling experience. Every detail of the bird is revealed and the cloud and sky form part of the picture, which is not possible with a telephoto lens.

COMMON NAME	Peregrine Falcon
SCIENTIFIC NAME	*Falco peregrinus*
DISTRIBUTION	Throughout Australia
SIZE	Female 40–50 cm, male 35–42 cm
BREEDING	2–4 eggs, use the same nest site (recess in cliff, hollow in large tree, abandoned large nests of other birds) year after year
DIET	Small to medium-sized birds, rabbits, other small mammals
STATUS	Moderately common

162

GALAH

Eolophus roseicapilla

Two parents feeding their young at the same time. After fledging from their nest hollows, young galahs form groups or creches, usually in the canopy of a tree. The parents go off to feed and return later. At this stage of their lives, the young are prone to attack by eagles and falcons. I have seen eight separate eagles' nests on one property, on the same day, with the ground below the nests covered in galah feathers. All had been harvested from these inexperienced young. Peregrines also catch many young galahs when they are in 'season' at this time. Galahs are one species that may have benefited from the clearing of native vegetation and the proliferation of cereals and pastures.

COMMON NAME	Galah
SCIENTIFIC NAME	*Eolophus roseicapilla*
DISTRIBUTION	Throughout Australia, except Cape York Peninsula, central Australia and inland from the Great Australian Bight
SIZE	35–36 cm
BREEDING	2–6 eggs, nests in hollow limb or hole in a tree
DIET	Fallen seeds
STATUS	Common

JUMPING SPIDER
Clynotus sp.

This spider is only 5 mm long. Highly magnified by the lens, it was photographed on a lerp-affected pink gum leaf. Jumping spiders are one of the dozens of species that make up the 'family tree' of creatures that depend on each species of plant in the bush. If the understorey is destroyed, trees are isolated, and may become more susceptible to attack by insects that can cause their death. The predators that control these insects need an all-year-round food supply, so they probably need a number of prey species, unlikely to be provided for by one type of tree.

COMMON NAME	Jumping Spider
SCIENTIFIC NAME	*Clynotus* sp.
SIZE	5–10 mm
BREEDING	10–20 eggs in a protective silken sac within a curled leaf or beneath tree bark
DIET	Crawling and flying insects
STATUS	Abundant

WHITE-BREASTED WOODSWALLOW

Artamus leucorynchus

This species of woodswallow's habit of sitting very close together on a branch when resting or roosting is very endearing. These small birds are very smartly turned out, when either perched or in flight, with their startling white under-wing and breast, and grey tail and head.

COMMON NAME	White-breasted Woodswallow
SCIENTIFIC NAME	*Artamus leucorynchus*
DISTRIBUTION	NT, QLD, NSW, inland VIC, northern SA, northern WA
SIZE	17 cm
BREEDING	3–4 eggs in nest in exposed part of tree
DIET	Insects
STATUS	Common to moderately common

167

KNOB-TAILED GECKO
Nephrurus laevissimus

What an amazing creature this is. I remember finding geckoes like this at a station when working on the dingo fence as a young jackeroo and marvelling at their unique colour and design.

This little fellow was found beside a road at night. The knob-tailed geckoes are reported to be voracious predators of other small lizards as well insects and spiders.

COMMON NAME	Knob-tailed Gecko
SCIENTIFIC NAME	*Nephrurus laevissimus*
DISTRIBUTION	Inland south-eastern WA, south-western NT, north-western SA
SIZE	6.5 cm
BREEDING	2 eggs
DIET	Crickets, cockroaches, beetles, spiders, scorpions, centipedes, caterpillars, other geckoes
STATUS	Common to very sparse

CHOCOLATE WATTLED BAT
Chalinolobus morio

After sunset on hot summer nights thousands of bats flock to waterholes and open water-tanks. They swoop across the water like nocturnal swallows and drop their bottom jaw in the water to collect a drink. Sometimes they mistime their approach and crash into the surface, then have to swim to the edge. Bats continue to drink in smaller numbers as the temperature drops during the night and sometimes a group will fly in just before dawn. It's a good idea to have a piece of wood floating in open-topped farm tanks to allow stranded bats, and birds, to escape.

COMMON NAME	Chocolate Wattled Bat
SCIENTIFIC NAME	*Chalinolobus morio*
DISTRIBUTION	South-eastern QLD, eastern NSW, VIC, eastern TAS, southern SA, southern WA
SIZE	Head and body length 5–6 cm, forearm length 3.5–4 cm
BREEDING	Young born in colonies, roost in tree hollows and roof cavities
DIET	Moths
STATUS	Common

LITTLE CORELLA
Cacatua sanguinea

A dead coolabah takes on a ghostly appearance with the approach of a thunderstorm. Although dead for many years, the tree provides a number of nest hollows for birds and bats and reptiles. Trees such as this, located near camp sites at outback waterholes, are being cut down for firewood and live trees often have every dead branch neatly cut off with a chainsaw.

COMMON NAME	Little Corella
SCIENTIFIC NAME	*Cacatua sanguinea*
DISTRIBUTION	North-western WA, north-eastern WA, NT, western QLD, western NSW, western VIC, eastern SA
SIZE	36–39 cm
BREEDING	2–3 eggs, nests in tree hollow
DIET	Seeds of grasses and legumes
STATUS	Common to locally common

WEDGE-TAILED EAGLE
Aquila audax

A young blond-headed wedge-tail is, by this stage, well experienced and a bold hunter in its own right. It stands in defiant, full threat display as other wedge-tails approach to share its kill. This is a common sight along outback highways, where thousands of kangaroos are killed each year by vehicles. Many kangaroos are drawn to the roadsides by the small green strip of feed that is the result of water run-off. As the eagles come in to feed on the road kill, they too are often injured by passing cars. Laden with food, an eagle is no match for a speeding car.

COMMON NAME	Wedge-tailed Eagle
SCIENTIFIC NAME	*Aquila audax*
DISTRIBUTION	Throughout Australia
SIZE	Female 89–104 cm, male 87–91 cm
BREEDING	1–3 eggs, nest is a large platform of sticks in a fork of a tree, re-used year after year
DIET	Rabbits, wallabies, small kangaroos, birds, reptiles
STATUS	Common

172

RAINBOW LORIKEET
Trichoglossus haematodus

A pair of lorikeets sitting among the foliage of a pink-flowering blue gum at Snelling's Beach on Kangaroo Island. These beautiful birds with their many colours are well camouflaged, but constantly advertise their presence with their rowdy calls. They can mimic the calls of other birds, such as noisy miners, and will join in with their varying alarm calls, reinforcing the sound.

COMMON NAME	Rainbow Lorikeet
SCIENTIFIC NAME	*Trichoglossus haematodus*
DISTRIBUTION	Kimberley region of WA to Cape York Peninsula, coastal QLD, NSW, VIC, to Eyre Peninsula in SA, around Perth, eastern TAS
SIZE	28 cm
BREEDING	2 eggs, nests in tree hollows
DIET	Pollen, nectar
STATUS	Abundant to common

GALAH
Eolophus roseicapilla

A small group of galahs in the top branches of a pink gum just after sunrise. This pair went through their usual ritual of preening and stretching. Some of the routine involved copying each other's movements. One scratched its head with its right foot and its mate used its left. One nibbled its left foot and the other its right. They looked like a mirror image of each other.

COMMON NAME	Galah
SCIENTIFIC NAME	*Eolophus roseicapilla*
DISTRIBUTION	Throughout Australia, except Cape York Peninsula, central Australia and inland from the Great Australian Bight
SIZE	35–36 cm
BREEDING	2–6 eggs, nests in hollow limb or hole in a tree
DIET	Fallen seeds
STATUS	Common

EMU

Dromaius novaehollandiae

After the opening rains in southern Australia, female emus begin to call. Their loud drumming calls can be heard for many kilometres at night, attracting males from some distance. As I watched this group, trying to sort out their territorial differences, they erupted in a chase of feathers and legs. Foxes sneak up on nesting emus at night and steal eggs. They can carry the giant eggs away in their huge jaws.

COMMON NAME	Emu
SCIENTIFIC NAME	*Dromaius novaehollandiae*
DISTRIBUTION	Throughout Australia except south coast of QLD and north coast of NSW, central VIC and south-eastern SA
SIZE	Up to 2 m
BREEDING	7–11 eggs in a grassy bed
DIET	Leaves, grasses, fruits, flowers, insects, seeds
STATUS	Locally abundant

COMMON WOMBAT
Vombatus ursinus

Venturing out in the evening the last thought on this fellow's mind was human technology. By this time I was tucked up in bed at home, or at least having dinner. In front of the wombat's burrow, I had left three flashlights and a camera attached to an electronic trigger, all mounted on a frame. As he emerged, he took his own photograph and vanished into the night.

COMMON NAME	Common Wombat
SCIENTIFIC NAME	*Vombatus ursinus*
DISTRIBUTION	Eastern NSW, VIC, TAS, south-eastern SA
SIZE	92–117 cm
BREEDING	1 young born, remains in the female's rear-opening pouch for about 6 months
DIET	Native grasses, sedges, matrushes, roots of shrubs and trees
STATUS	Common

YELLOW-FOOTED ROCK-WALLABY
Petrogale xanthopus

On frosty spring mornings these rock-wallabies greet the day sitting in the first rays of sun to warm their bodies. If a cold wind is blowing from the east they prefer to sit in the lee of a hillside and forgo the sunshine.

Surveys carried out by the National Parks and Wildlife Service use this behaviour to advantage when carrying out aerial counts as the wallabies are easy to spot at this time of day.

COMMON NAME	Yellow-footed Rock-wallaby
SCIENTIFIC NAME	*Petrogale xanthopus*
DISTRIBUTION	Flinders Ranges, SA, Barrier Range, NSW, Grey Range, QLD
SIZE	Head and body length 48–65 cm, tail length 57–70 cm
BREEDING	1 young
DIET	Grasses, plants
STATUS	Common in Flinders Ranges, rare elsewhere